AF478645

Villa Farnese
Roma 13 Dicembre 2002
A. Lavi

Trends in Mathematics

Trends in Mathematics is a book series devoted to focused collections of articles arising from conferences, workshops or series of lectures.

Topics in a volume may concentrate on a particular area of mathematics, or may encompass a broad range of related subject matter. The purpose of this series is both progressive and archival, a context in which to make current developments available rapidly to the community as well as to embed them in a recognizable and accessible way.

Volumes of TIMS must be of high scientific quality. Articles without proofs, or which do not contain significantly new results, are not appropriate. High quality survey papers, however, are welcome. Contributions must be submitted to peer review in a process that emulates the best journal procedures, and must be edited for correct use of language. As a rule, the language will be English, but selective exceptions may be made. Articles should conform to the highest standards of bibliographic reference and attribution.

The organizers or editors of each volume are expected to deliver manuscripts in a form that is essentially "ready for reproduction." It is preferable that papers be submitted in one of the various forms of LaTeX in order to achieve a uniform and readable appearance. Ideally, volumes should not exceed 350-400 pages in length.

Proposals to the Publisher are welcomed at either:
Birkhäuser Boston, 675 Massachusetts Avenue, Cambridge, MA 02139, U.S.A.
kostant@birkhauser.com
or
Birkhäuser Publishing, Ltd., 40-44 Viaduktstrasse, CH-4051 Basel, Switzerland
hempfling@birkhauser.ch

Multiscale Methods in Quantum Mechanics

Theory and Experiment

Philippe Blanchard
Gianfausto Dell'Antonio

Editors

Birkhäuser
Boston • Basel • Berlin

Philippe Blanchard
Universität Bielefeld
Fakultät für Physik
33615 Bielefeld
Germany

Gianfausto Dell'Antonio
Università degli Studi di Roma
 "La Sapienza"
Dipartimento di Matematica
00185 Roma
Italy

Library of Congress Cataloging-in-Publication Data
Multiscale methods in quantum mechanics : theory and experiment / Philippe Blanchard,
    Gianfausto Dell'Antonio, editors.
        p. cm. – (Trends in Mathematics)
    Includes bibliographical references.
    ISBN 0-8176-3256-5 (alk. paper)
    1. Quantum theory. I. Blanchard, Philippe. II. Dell'Antonio, G. F., 1933- III.
Series.

QC174.12.M85 2004
530.12–dc22   2004045074
                                                                  CIP

ISBN 0-8176-3256-5   Printed on acid-free paper.

Printed in the United States of America.   (TXQ/MV)

9 8 7 6 5 4 3 2 1   SPIN 10934654

Birkhäuser is a part of *Springer Science+Business Media*

www.birkhauser.com

Preface

This volume contains papers which were presented at a meeting entitled "Multiscale Methods in Quantum Mechanics: Theory and Experiment," held at the Academia dei Lincei in Rome (December 16–20, 2002). The organizing and scientific committee was composed of Ph. Blanchard (University of Bielefeld), G.F. Dell'Antonio (University of Rome 1), S. Graffi (University of Bologna), and Th. Paul (Ecole Normale Supérieure, Paris).

The conference could not have happened without the generous support of numerous sponsors: the Centro Linceo Interdisciplinare of the Academia dei Lincei in Rome, the Istituto di Alta Matematica of the CNR through the National Group of Mathematical Physics and the Mathematical Physics Sector of the SISSA in Trieste. We are most grateful to all of them.

The central theme of the meeting was the comparison of different techniques and strategies used in mathematical physics to tackle the challenging problems in quantum mechanics of treating the fast and slow degrees of freedom and other multiscale phenomena. In classical mechanics the treatment of the corresponding problem is based on various versions of the "adiabatic theorem," which requires substantial modifications in order to be applicable to quantum mechanics. The topic of multiscale decomposition has attracted much attention recently from research groups in mathematical physics, in part due to the increasing number of significant physical applications, from the refinement of the Born–Oppenheimer approximation to the study of the motion of charged particles in slowly varying electromagnetic fields up to decoherence. Multiscale methods have also been applied successfully in the analysis of the propagation of electromagnetic waves in media with slowly varying indexes of refraction.

Different extensions of adiabatic-like methods motivated by quantum mechanics have been proposed. They share on the one hand some common elements but differ on the other hand depending on their different emphases, mathematical techniques used, and their results. The goal is to provide a better synthesis and contribute therefore to the development of this very interesting and fast developing field of research. In view of the interdisciplinary nature of the subject, each day a lecture was devoted to an experimental topic. Indeed, the central theme of this book is a very good example of the way in which mathematics and physics are complementary. Displaying and encouraging this interaction, one hopes that further insight will arise from the cross-fertilization of the two cultures, as has already been the case in the past.

The success of the conference was due first of all to the speakers. Thanks to their efforts, it was possible to take into account recent developments as well as open problems and to make the conference an exciting meeting. We hope that participants and readers will find these articles interesting and useful.

It is a pleasure to express our gratitude to Anna Anastasi for assistance before and during the meeting and Hanne Litschewsky for help in preparing the conference and in collecting and editing the manuscripts for publication.

Ph. Blanchard and G.F. Dell'Antonio

Bielefeld, Trieste, November 2003

Contents

*Multiscale Methods
in Quantum Mechanics*

Organic Molecules and Decoherence Experiments in a Molecule Interferometer

M. Arndt, L. Hackermüller, K. Hornberger, and A. Zeilinger

Institut für Experimentalphysik
Universität Wien
Boltzmanngasse 5
A-1090 Wien, Austria

1.1 Classical behaviour from quantum physics

One of the basic objectives in the foundations of physics is to understand the detailed circumstances of the transition from pure quantum effects to classical appearances. This field has gained an even increased attention because of the fact that quantum phenomena on the mesoscopic or even macroscopic scale promise to be useful for future technologies, such as in lithography with clusters and molecules, quantum computing or highly sensitive quantum sensors.

In the present work we describe one specific type of quantum effect, namely the wave-particle duality and the requirements to observe this phenomenon. The wave-nature of matter has been known for the last 80 years and it finds daily technological confirmation and application in the material sciences which use electron diffraction, electron holography or neutron diffraction to investigate surfaces or bulk material.

Here we shall focus on the extension of matter-wave experiments towards more massive, more complex and larger quanta, ranging from carbon clusters, over biologically relevant organic molecules to fullerene derivatives.

The non-observability of quantum wave effects in our daily world can be described by various complementing arguments: The first set of theories assumes an objective reduction of the wavefunction, which may be described by a non-linear extension of the Schrödinger equation [8, 18]. The second set of theories takes the unitarity of the Schrödinger equation as the central element of quantum physics and tries to explain the appearance of classical phenomena within the well-established linear dynamics.

The simplest argument of this category is that quantum effects are often simply too small to be observable. For instance, any diffraction pattern of a human being would take a diffraction experiment larger than the size of the known universe to separate the interference fringes to a measurable distance. The de Broglie wavelength of a walking human being is actually below the

Planck length of 10^{-35} m — and one can currently only speculate about the meaning of any wave physics with dimensions below that scale.

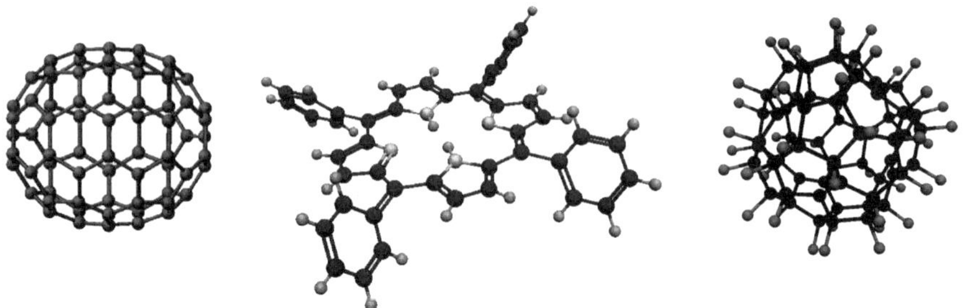

Fig. 1.1. The fullerene C_{70} is composed of 70 carbon atoms in an oval shape (*left*). Tetraphenylporphyrin $C_{44}H_{30}N_4$ (TPP, m=614 amu) is composed of four tilted phenyl rings attached to a planar porphyrin structure (*middle*). The largest dimension measures about 2 nm and the aspect ratio (height to width) is roughly seven. The fluorinated fullerene $C_{60}F_{48}$ (m=1632 amu) (*right*) is a deformed bucky-ball surrounded by a shell of 48 fluorine atoms. This molecule exists in different isomers and only one structure with D_2 symmetry is shown here [20].

Now, while this dynamical argument certainly holds, it seems there is another mechanism which limits the appearance of quantum effects even for smaller objects under real-world conditions: Decoherence theory, which started to flourish about two decades ago [21, 22, 14, 9, 2, 23], states that isolated quantum systems may always maintain their quantum nature — independent of their size or mass — but that it becomes increasingly difficult to guarantee this isolation for large objects. The qualitative idea is that any coupling between the system and the environment will lead to an entanglement of the two. Any coherence in the quantum system will be rapidly diffused into the complex environment and will therefore no longer be observable. As soon as the environment can distinguish between different quantum states of the system, i.e., as soon as which-state information becomes accessible to the outer world, a superposition of these states is no longer observable.

A simple example is given by the fact that a human being is in perpetual interaction with its environment through collisions with the surrounding air molecules. It is very hard to quantitatively follow the loss of coherence in such a macroscopic body, since about 10^{27} molecules impinge on a human body per second. Decoherence is thus expected to occur on time scales well below any available experimental resolution.

However, as we shall discuss below, macromolecule interferometry allows us to study the effect of collisions quantitatively as one important origin of the transition from the quantum to the classical: Molecules in a high-vacuum environment exhibit pure quantum properties if properly prepared. Since the background gas pressure may be well controlled by the experimentalist, quantum effects may be turned gradually and in a controlled way into classical appearances.

1.2 Coherence experiments

The quantum wave nature of material objects is well established for small particles. But the conceptual dissonance between the local character of the 'particle' and the non-local character of the 'wave' seems to strike our common sense even stronger if we consider objects which can already be clearly seen as isolated particles under a microscope. This is the case for nanometer sized molecules and clusters like the fullerenes, porphyrins or fullerene derivatives which are shown in Fig. 1.1 (left to right).

Fullerenes, named after the geodesic domes created by the architect Buckminster Fuller [16], were discovered only in 1985 by Kroto and coworkers [15]. The 'buckyball' C_{60} and the 'bucky-rugbyball' C_{70} (Fig. 1.1, left) were the first macromolecules for which wave-particle duality was demonstrated [1] mainly because they can be easily vaporized, they are very stable, and they can be efficiently detected using laser ionization [17]. In addition to these practical reasons they offer many similarities to bulk material: They possess bulk-like excitations such as phonons, excitons and plasmons. They may be regarded as their own internal heat bath, and they are known to emit thermal electrons or thermal photons or even diatomic molecules when they are heated to sufficiently high temperatures. In that sense they are a very good approximation to a classical body on the nanoscale. They have a mass of 840 amu and a diameter of roughly 1 nm. They are rigid and highly symmetric.

In contrast to the fullerenes the second species used in our experiments, the porphyrins, are very abundant in nature. The porphyrin structure can be found in the core of a heme molecule or in chlorophyll and since a metal atom inside the porphyrin structure gives blood its red color and chlorophyll its green appearance, porphyrins are often referred to as the 'colors of life'. The particles used in our experiments are derivatives of this porphyrin structure: They are enlarged by four tilted phenyl rings, which lead to the full name tetraphenylporphyrin (TPP, Fig. 1.1, center). Although even successful interference experiments with porphyrins are obviously far from proving the relevance of quantum mechanics in the living world, they are nevertheless an interesting twist in an experiment on the foundations of quantum physics: The diameter of the TPP molecule — measured by the core to core distance of the outermost nuclei — is three times as large as that of C_{70}, it is flat instead of nearly spherical, it has dangling phenyl rings sticking out and the polarizability is much more anisotropic than in the case of the fullerenes.

Finally, we have also performed matter wave interferometry with the fluorinated fullerene $C_{60}F_{48}$. It has the shape of a deformed buckyball with 48 fluorine atoms covalently bound to the outer shell. The diameter of this molecule is in between that of the pure buckyball and the porphyrins, but the mass of 1632 amu is about twice the mass of C_{70}. It represents the current record in mass, complexity and number of atoms contained in a single quantum object in a matter wave experiment.[1] The fluorinated fullerenes exist in several isomers of different symmetry which were probably present in all configurations.

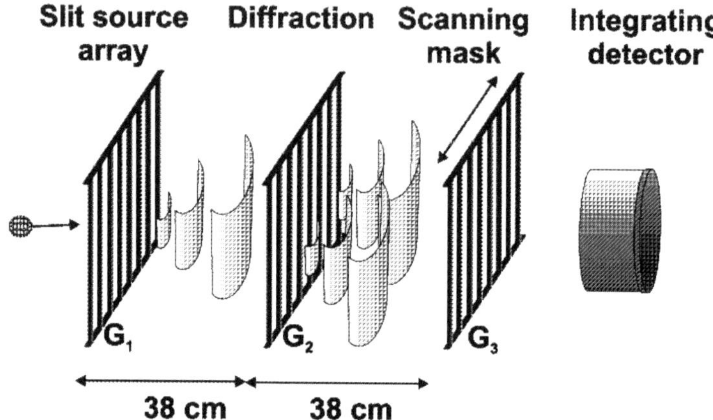

Fig. 1.2. Setup of the Talbot–Lau interferometer: hot, thermal molecules enter a near-field interferometer, which consists of three identical gold gratings. The first grating represents an array of slit sources which imprint the required coherence onto the uncollimated, spatially incoherent molecular beam. The second grating images this slit source onto the plane of the third grating. The third grating serves as a mask for the detection of the interference fringes: Only molecules position-synchronized with the grating will pass the structure and may be observed in the following detector, composed of laser ionization and ion counting [5].

The best way to prove an assumed wave-like phenomenon is to observe interference. Although in earlier experiments with C_{60} molecules far-field diffraction was an ideal tool for this purpose because of its conceptual simplicity, all experiments described here were performed in a near-field interferometer of the Talbot–Lau type which has been proposed by Clauser et al. [7, 6] and which has only recently been applied to large molecules [5, 4]. There are two main reasons for this choice: Firstly, the three grating arrangement allows us to work with an uncollimated, spatially incoherent molecule source. This is essential for the biomolecules and large clusters, since typical source intensities are much below those of atomic or fullerene beams. Secondly, the

---

[1] It is important not to confuse the macroscopically occupied field of a Bose–Einstein condensate (BEC) with the single many-body wavefunction of a complex molecule: the de Broglie wavelength of a BEC is always that of a single atom, while the molecular λ_{dB} is determined by all participating atoms.

dimension requirements (grating constant, and grating separation) are less severe in a near-field interferometer compared to far-field diffraction or far-field interferometry. This is seen by the fact that the grating constant in our present near-field interferometer is 1μm, while it had to be ten times smaller for the previous far-field experiments [1].

The experimental setup is shown in Fig. 1.2. Molecules are sublimated in an oven at the maximum allowed temperature before the onset of molecular decomposition, i.e., at $890\,$K for C_{70}, $690\,$K for TPP and $560\,$K for $C_{60}F_{48}$. The molecules fly into the high vacuum chamber and traverse the interferometer, before they enter the detector. In all C_{70} experiments the detector consists of a laser ionization stage and subsequent ion counting. This method is very efficient ($\eta_{\mathrm{ion}} \sim 10\%$) [17] but highly molecule specific. Most large organic molecules would rather fragment than ionize. For the porphyrins and fluorofullerenes we therefore employ electron impact ionization, which also leads to the formation of positive ions. However, the efficiency is reduced to $\eta_{ion} \sim 10^{-4}$. Since electron impact is not molecule selective we add a quadrupole mass selection stage to separate the porphyrins and fluorofullerenes from air molecules and contaminations in the vacuum chamber.

The interferometer itself consists of three gold gratings, with a grating constant of $990\,$nm and an open fraction of about $f = 0.48$, i.e., open gaps of about $470\,$nm. The gratings are separated by a distance of $L = 38\,$cm [11].

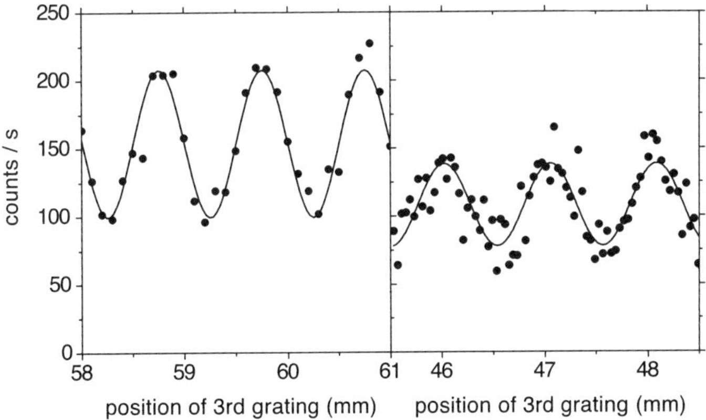

Fig. 1.3. Left: Interference pattern recorded for porphyrin molecules at 166 m/s. The interference contrast of 38% is in good agreement with quantum mechanical expectation and in clear disagreement with the classical value of 13%. The interferogram was recorded over a few minutes. Right: The same experiment performed with $C_{60}F_{48}$ and recorded over three hours. The picture consists of an average of 13 'low-noise' single scans which were selected by the magnitude of their frequency noise (not by the value of the frequency). [10].

The longitudinal, i.e., spectral coherence length $l_c = \lambda^2/\Delta\lambda$ inside the oven is determined by the velocity spread in the source and it is only about 1.5 times larger than the de Broglie wavelength, which lies for all particles and velocities in the range of $2\ldots5 \times 10^{-12}$ m. Both the de Broglie wavelength and the coherence length are thus $500\ldots1000$ times smaller than the size of the molecules themselves! The transverse coherence is equally small when the molecules leave the oven, and in that sense it is fully justified to speak of well-localized particles in the preparation stage. However, each slit in the first grating boosts the transverse coherence length of the molecular beam at the position of the second grating by about six orders of magnitude [3]. The indistinguishable passage through two or more slits leads to quantum interference behind the second grating, in particular to the formation of a molecular density pattern with the same period as the diffraction structure in about the distance of the Talbot length $L_T = g^2/\lambda_{dB}$ behind the grating. Contributions from different slits in the first grating add up incoherently but in phase. The molecular density distribution can be probed using a third grating with again the same period, which is shifted transversely to the molecular beam. Whenever the molecular structure and the gold grating are in phase, most molecules will pass it and reach the detector. If the grating is slightly shifted, the count rate is reduced. The interference pattern is thus revealed by scanning the mask across the molecule beam.

Fig. 1.3 (left) shows typical high-contrast interference fringes of porphyrin. The contrast of 38% is in good agreement with the theoretically expected value and it is about a factor of three higher than the classically expected shadow contrast. Both the classical and the quantum calculation include the van der Waals interaction between the molecules and the 500 nm thin grating walls [11]. The quantum mechanical wave nature can further be proved by the way the interference pattern changes with the de Broglie wavelength: We can vary the molecule velocity and thus de Broglie wavelength. Since the Talbot length depends on λ_{dB}, also the interference contrast is effected by the varying velocity and we find very good agreement between theory and experiment for all velocity classes (see [11]).

For the fluorinated fullerenes the same experiment was repeated. Fig. 1.3 (right) shows the result for $C_{60}F_{48}$. The contrast of 27% is clearly above the classical expectation of 12% but still below the best possible quantum interference contrast of 37%. We attribute this difference mainly to vibrations on the level of a few ten nanometers which affect slow molecules more than fast ones. $C_{60}F_{48}$ had a most probable speed of 100 m/s while the TPP interference pattern was recorded for a mean molecular velocity of 166 m/s. The vibration hypothesis is further supported by the fact that slow fullerenes can also show a decrease in fringe visibility by about 30% and also by dedicated vibration experiments [19]. The second reason is the large technical background noise in these experiments in addition to the low count rates and correspondingly long integration times. Both effects tend to reduce the interference contrast.

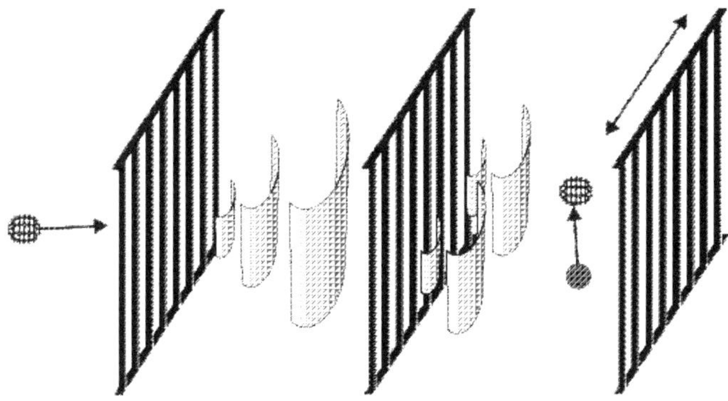

Fig. 1.4. Idea of the decoherence experiment: The originally delocalized molecules may be well localized through interactions with the environment. Collisions with background gas molecules start to become relevant at pressures above 10^{-7} mbar. The gas pressure representing the environment can be well controlled by the experimentalist.

1.3 Decoherence experiments

Having shown a high level of coherence in experiments with large molecules, it is now interesting to investigate the limit of matter wave interferometry due to the coupling between the quantum object and its environment. These studies have again been performed with C_{70} to make use of the much higher count rates and better statics compared to the TPP and $C_{60}F_{48}$ experiments.

One can imagine many different couplings between a particle and its environment. Collisions and radiative interactions will be the most frequent and most natural processes in our macroworld, and here we shall focus on collisions in particular. A fullerene molecule in the superposition of two position eigenstates $\left|C_{70}^{\text{left}}\right\rangle, \left|C_{70}^{\text{right}}\right\rangle$ will get entangled with a gaseous environment, since the state of an incident gas particle $|g\rangle$ will be changed depending on the position of the fullerenes:

$$|\psi\rangle = \frac{1}{\sqrt{2}}\left[\left|C_{70}^{\text{left}}\right\rangle + \left|C_{70}^{\text{right}}\right\rangle\right] \otimes |g\rangle \xrightarrow{\text{coll.}} \frac{1}{\sqrt{2}}\left[\left|C_{70}^{\text{left}}\right\rangle\left|g_{\text{scat}}^{\text{left}}\right\rangle + \left|C_{70}^{\text{right}}\right\rangle\left|g_{\text{scat}}^{\text{right}}\right\rangle\right]$$

Any 'measurement' on the state of the scattered gas particle, by further interactions with other gas molecules or the walls, can be regarded as an effective measurement of the fullerene path. If the scattered gas states are distinguishable, i.e., $\left\langle g_{\text{scat}}^{\text{left}} \middle| g_{\text{scat}}^{\text{right}} \right\rangle = 0$, interference between $\left|C_{70}^{\text{left}}\right\rangle$ and $\left|C_{70}^{\text{right}}\right\rangle$ can no longer be observed and the superposition state $|\psi\rangle$ has effectively collapsed.

A more detailed description of the collisional decoherence rate in collision experiments has been given in [13, 12]. The main result with respect to our investigation is that we expect an exponential loss of the interference contrast with increasing pressure of the residual gas

$$V(p) = V_0 \exp\left(-\frac{2L\sigma_{\text{eff}}}{k_B T}p\right) = V_0 \exp(-p/p_v). \qquad (1.1)$$

In Fig. 1.5 we show the experimental result for fullerenes in interaction with Argon atoms at base pressures between $10^{-8} \ldots 10^{-6}$ mbar. We find the predicted single-exponential decay, and the theoretical decay curve fits the experiment without free parameter, except for the contrast at $p = 0$ mbar, which is also influenced by vibrations or other effects that are independent of the base pressure. We emphasize that the exponential decay in Fig. 1.5 is a clear signature of decoherence. In the case of simple absorption it would be the beam intensity that shows an exponential decrease, while the visibility would remain constant. Note also that decoherence theory describes the experiment not only qualitatively but quantitatively. The experiment allowed even falsification of an incorrect localization rate that differs by a factor of 2π (see the discussion in [12]).

Let us note that there are two equivalent ways of viewing the decoherence process observed in the experiment. On the one hand one can say that the change in the state of the scattered gas particle is strong enough to obtain full information on the position of a fullerene by a single collision. On the other hand one can view the fullerene in the momentum representation. Then the recoil during a single collision translates the momentum distribution such that the final interference pattern is shifted substantially with respect to the unscattered case. Since the collision angles turn from a quantum superposition into a broad distribution once the gas particle is 'measured' by the environment, the summation of many randomly shifted single-molecule interference patterns again washes out the total fringe pattern. It is worth noting that in our experiment a sizeable number of the collisions between fullerenes and residual gas particles lead to kicks out of the detection range and therefore to loss. These events do not contribute to the reduction of the fringe contrast. Only small angle collisions, with $\theta \sim 1$ mrad, leave the molecules inside the interferometer while they still dephase the interference fringes. We can extrapolate the numerical data which we obtained in similar decoherence experiments with different residual gases to predict the feasibility of experiments with even larger objects than the porphyrins or fluorofullerenes. It turns out that even for objects with masses around 10^7 amu the $1/e$-decoherence pressure will 'only' be $\sim 10^{-10}$ mbar, a value which can still be reached with standard laboratory equipment [10]. Collisional decoherence — although very efficient on the macro-scale — is thus still far from being a limit to interferometry for objects with sizes below 10 nm. However, the influence of radiative interactions, of quasi-static interactions with fluctuating or strongly dispersive potentials for molecules with a large electric polarizability, electric or magnetic dipole moment still remains to be investigated.

Quantum optics with macromolecules has only begun. We foresee many important developments and future improvements concerning the sources for neutral macromolecules, more efficient detection schemes and novel types of

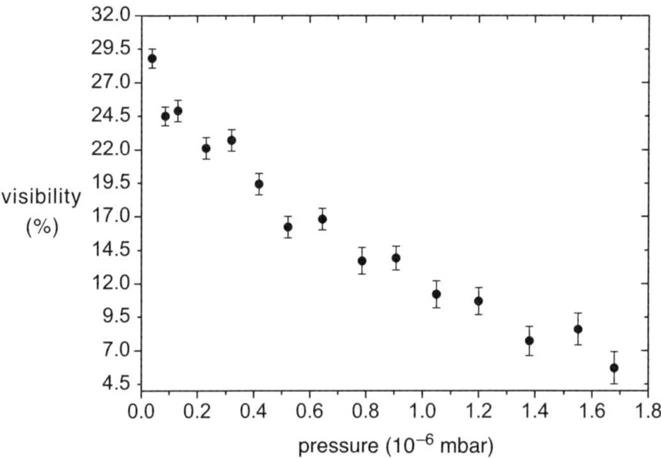

Fig. 1.5. Interference fringe visibility for C_{70} fullerenes interacting with Argon gas at various pressures in the interferometer. A clear single-exponential decay is observed as expected from decoherence theory.

interferometers. Applications will range from fundamental studies of decoherence to the potential use of molecule lithography and nanotechnology.

Acknowledgements. We thank our colleagues Björn Brezger, Elisabeth Reiger and Stefan Uttenthaler, for their contribution to the described experiments. This work has been supported by the Austrian Science Foundation (FWF), within the project START Y177 and SFB F1505, as well as by the European networks HPRN-CT-2000-00125, HPRN-CT-2002-00309.

References

1. M. ARNDT, O. NAIRZ, J. VOSS-ANDREAE, C. KELLER, G. V. DER ZOUW AND A. ZEILINGER, Wave-particle duality of C_{60} molecules, *Nature*, **401** (1999), 680–682.
2. P. BLANCHARD, D. GIULINI, E. JOOS, C. KIEFER AND I.-O. STAMATESCU, HRSG., *Decoherence: Theoretical, Experimental, and Conceptual Problems*, Vol. 538, Lecture Notes in Physics, Springer-Verlag, Berlin, 2000.
3. M. BORN AND E. WOLF, *Principles of Optics*, Pergamon Press, 1993.
4. B. BREZGER, M. ARNDT AND A. ZEILINGER, Concepts for near-field interferometers with large molecules, *J. Opt. B: Quantum Semiclass. Opt.*, **5** (2003), S82–S89.
5. B. BREZGER, L. HACKERMÜLLER, S. UTTENTHALER, J. PETSCHINKA, M. ARNDT AND A. ZEILINGER, Matter-wave interferometer for large molecules, *Phys. Rev. Lett.*, **88** (2002), 100404.
6. J. CLAUSER AND S. LI, Generalized Talbot–Lau atom interferometry, in *Atom Interferometry*, P. R. Berman, Hrsg., Bd. 11, Academic Press, 1997, 121–151.

7. J. CLAUSER AND M. REINSCH, *New theoretical and experimental results in fresnel optics with applications to matter-wave and x-ray interferometry*, Applied Physics B, **54** (1992), 380.

8. G. C. GHIRARDI, A. RIMINI AND T. WEBER, *Unified dynamics for microscopic and macroscopic systems*, Phys. Rev. D, **34** (1986), 470–491.

9. D. GIULINI, E. JOOS, C. KIEFER, J.KUPSCH, O. STAMATESCU AND H. D. ZEH, *Decoherence and the Appearance of the Classical World in Quantum Theory*, Springer-Verlag, Berlin, 1996.

10. L. HACKERMÜLLER, K. HORNBERGER, B. BREZGER, A. ZEILINGER AND M. ARNDT, *Decoherence in a Talbot Lau interferometer: The influence of molecular scattering*, Applied Physics B, **77** (2003), 781–787.

11. L. HACKERMÜLLER, S. UTTENTHALER, K. HORNBERGER, E. REICER, B. BREZGER, A. ZEILINGER AND M. ARNDT, *Wave nature of biomolecules and fluorofullerenes*, Phys. Rev. Lett., **91** (2003), 90408.

12. K. HORNBERGER AND J. E. SIPE, *Collisional decoherence reexamined*, Phys. Rev. A, **68** (2003), 012105.

13. K. HORNBERGER, S. UTTENTHALER, B. BREZGER, L. HACKERMÜLLER, M. ARNDT AND A. ZEILINGER, *Collisional decoherence observed in matter wave interferometry*, Phys. Rev. Lett., **90** (2003), 160401.

14. E. JOOS AND H. D. ZEH, *The emergence of classical properties through interaction with the environment*, Z. Phys. B., **59** (1985), 223–243.

15. H. KROTO, J. HEATH, S. O'BRIAN, R. CURL AND R. SMALLEY, C_{60}: *Buckminsterfullerene*, Nature, **318** (1985), 162–166.

16. R. W. MARKS, *The Dymaxion World of Buckminster Fuller*, Reinhold, New York, 1960.

17. O. NAIRZ, B. BREZGER, M. ARNDT AND A. ZEILINGER, *Diffraction of complex molecules by structures made of light*, Phys. Rev. Lett., **87** (2001), 160401-1.

18. R. PENROSE, *On gravity's role in quantum state reduction*, Gen. Rel. Grav., **28** (1996), 581–600.

19. A. STIBOR ET AL. in preparation, 2003.

20. S. I. TROYANOV, P. A. TROSHIN, O. V. BOLTALINA, I. N. IOFFE, L. N. SIDOROV AND E. KEMNITZ, *Two isomers of $C_{60}F_{48}$: An indented fullerene*, Angew. Chem. Int. Ed., **40** (2001), 2285.

21. H. D. ZEH, *On the interpretation of measurement in quantum theory*, Found. Phys., **1** (1970), 69–76.

22. W. H. ZUREK, *Pointer basis of quantum apparatus: Into what mixture does the wave packet collapse?*, Phys. Rev. D, **24** (1981), 1516–1525.

23. W. H. ZUREK, *Decoherence, einselection, and the quantum origins of the classical*, Rev. Mod. Phys., **75** (2003), 715–775.

Colored Hofstadter Butterflies

J.E. Avron

Department of Physics
Technion
32000 Haifa, Israel

Summary. I explain the thermodynamic significance, the duality and open problems associated with the two colored butterflies shown in Figures 2.1 and 2.4.

2.1 Overview

My aim is to explain what is known about the thermodynamic significance of the two colored butterflies shown in Figures 2.1 and 2.4 and what remains open. Both diagrams were made by my student, D. Osadchy [14], as part of his M.Sc. thesis. I shall explain their interpretation as the $T = 0$ phase diagrams of a two-dimensional gas of charged, though non-interacting, fermions. Fig. 2.1 is associated with weak magnetic fields (and strong periodic potentials) while Fig. 2.4 with strong magnetic fields (and weak periodic potentials). The two cases are related by duality. The duality, which is further discussed below, is manifest if colors are disregarded.

The horizontal coordinate in both figures is the chemical potential μ and the vertical coordinate is proportional to the magnetic induction B in Fig. 2.1 and $1/B$ in Fig. 2.4. The colors represent the quantized values of the Hall conductance, i.e., represent integers.[1] Warm colors represent positive multiples and cold colors represent negative ones: Orange represents 2, red 1, white 0, blue -1 etc.

Remark: *It is problematic to represent integers by colors with good contrast between nearby integers. This is related to the fact that colors are not ordered on the line but rather are represented by the simplex (r, g, b) with $r + g + b = 1$. (Pure colors are located on the boundary of the simplex.) The assignment in the figures becomes problematic for large, positive or negative, integers: Large positive integers are not represented anymore by warm colors but rather by yellow and green.*

---

[1] The quantum unit of conductance, e^2/h, is $1/2\pi$, in natural units where $e = \hbar = 1$.

I shall also present an open problem. Namely, how do these diagrams change if one replaces the magnetic induction B by the magnetic field H as the thermodynamic coordinate.

2.2 Some history

That the Hall conductance took different signs in different metals was an embarrassment to Sommerfeld theory. Since charge is carried by the electrons, one sign was predicted. The wrong sign was called the anomalous Hall effect and was explained by R. Peierls [5] who showed that the periodicity of the electron dispersion $\epsilon(k)$ plus the Pauli principle allow for either sign, depending on μ. This subsequently led to the important concept of holes as charge carriers—a term not used by Peierls in his original work.

The electron-hole anti-symmetry of the Hall conductance is seen in Fig. 2.1 where cold and warm colors are interchanged upon reflection about the vertical axis. However, the figure is much more complicated than what Peierls had in mind.

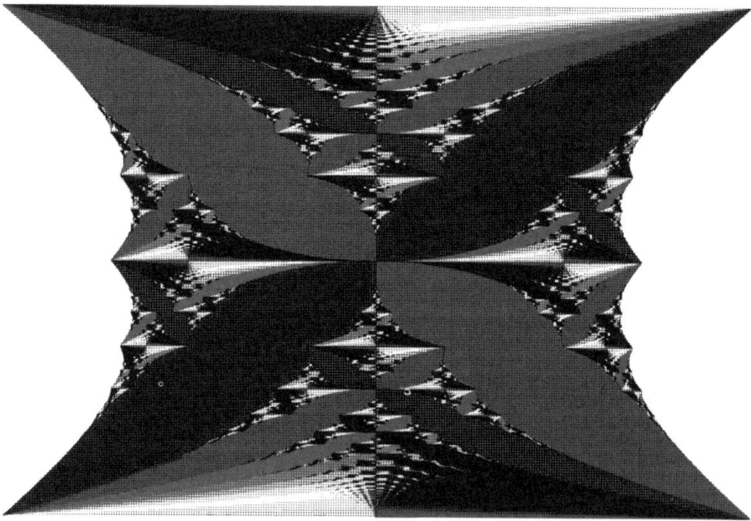

Fig. 2.1. Colored Hofstadter butterfly for Bloch electrons in weak magnetic field. The horizontal axis is the chemical potential; the vertical axis is the magnetic flux through the unit cell. The diagram is periodic in the flux and one period is shown. It admits a thermodynamic interpretation of a phase diagram. (See color insert following p. 22.)

Mark Azbel [2] realized that the Schrödinger equation in a periodic potential and magnetic field had tantalizing spectral properties. But it was the graphic rendering of the spectrum by D. Hofstadter [9], shown in Fig. 2.2, (and his scaling rules,) that brought the problem into the limelight. The richness of spectral properties is a result of competing area scales: One dictated by the unit cell of the underlying periodic potential and the other by the area that carries one unit of magnetic flux. When Φ is rational the two areas are commensurate, when it is irrational, they are not. At $T = 0$ the electron gas is coherent on large distance scales and commensuration leads to interference phenomena that affect spectral properties at very small energy scales. The delicate spectral properties attracted considerable attention of a community of spectral analysts. Reference [1] is a pointer to a rich and wonderful literature on the subject.

Fig. 2.2. The original, monochrome, Hofstadter butterfly, shows the spectrum, on the horizontal axis, as function of the flux Φ which is the vertical axis. The spectrum is the complement of the colored set shown in Fig. 2.1.

In a seminal work, TKNN [18] realized that the Hall conductance of the Hofstadter model admits a topological characterization in terms of Chern numbers. This discovery is an interesting piece of lost history. In fact S. Novikov was apparently the first to realize the topological significance of the spectral gaps for Bloch electrons in magnetic fields [13]. However, he missed their significance as Hall conductance. TKNN [18] were not aware of the work of Novikov. Instead, they were motivated by a puzzle that follows from applying the Laughlin argument for the quantization of the Hall conductance to the Hofstadter model. By essentially reinventing the proof of integrality of Chern numbers in a special case, they showed that whenever the Fermi energy is in a gap the Hall conductance is quantized.

In the two diagrams, Figs. 2.1 and 2.4, the Hall conductance is quantized almost everywhere. The set of points where the Hall conductance is not quantized is a set of zero measure and so invisible. This is related to the fact that the spectrum is a set of measure zero (see e.g., [1] and references therein).

2.3 Thermodynamics considerations

It is interesting to consider the colored butterflies from the perspective of thermodynamics.

2.3.1 Gibbs phase rule

The first and second laws of thermodynamics constrain the shape of phase diagrams. The phase rules depend on the choice of the independent thermodynamic coordinates X, be they extensive, such as $X = (E, V, N)$, or intensive, such as (P, T).

Let $X = (E, V, N)$ be the extensive coordinates of a simple thermodynamic system. X and λX with $\lambda > 0$ are thermodynamically equivalent systems, while X and $Y \neq \lambda X$ are not. Mixing X and Y, in any proportion, is, in general, an irreversible process. The second law then says that the entropy of the mixed system is not smaller than the sum of the entropies of its constituents. Namely, for $0 \leq \lambda \leq 1$,

$$ S(\lambda X + \lambda' Y) \geq S(\lambda X) + S(\lambda' Y) = \lambda S(X) + \lambda' S(Y), \quad \lambda' = 1 - \lambda. \quad (2.1) $$

The first law, conservation of energy, (plus conservation of number of particles and additivity of volumes), was used in the first step and the extensivity of the entropy, $S(\lambda X) = \lambda S(X)$, in the second. Eq. (2.1) says that the entropy $S(X)$ is a concave function of its arguments. This embodies the basic laws of thermodynamics.

Equality in Eq. (2.1) holds if mixing is reversible which is, of course, the case if a phase is mixed with itself. It is also the case if coexisting phases are mixed: Clearly one can separate ice from water by mechanical means alone. The geometric expression of equality in Eq. (2.1) is that S contains a linear segment: For a pure phase this is the half line $S(\lambda X) = \lambda S(X)$. When $X \neq \lambda Y$ are in coexistence, S contains a two-dimensional cone: $S(\lambda_1 X + \lambda_2 Y) = \lambda_1 S(X) + \lambda_2 S(Y)$, $\lambda_{12} > 0$. (This notion extends to multiple phase coexistence.)

Positivity of the temperature implies that S is an increasing function of E. Consequently, $S(E, V, N)$ can be inverted to give the internal energy $E(S, V, N)$. Since S is a concave function of its arguments, $E(S, V, N)$ is a convex function of its arguments (which are the extensive state variables). Its

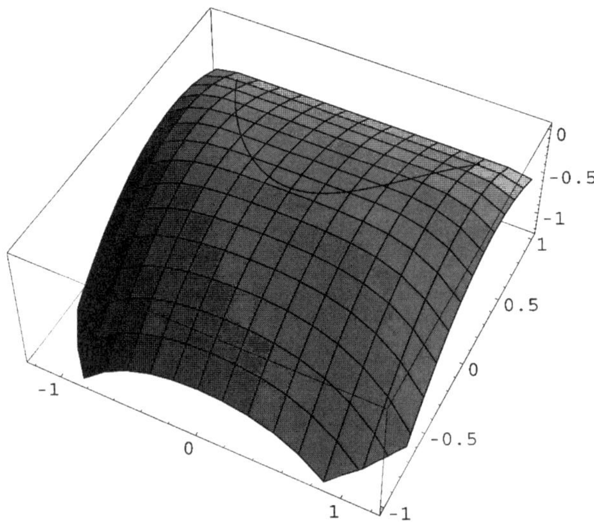

Fig. 2.3. $S(E, V, N)$ is a concave function shown here for N fixed. The strictly convex pieces are associated with pure phases. The ruled piece is where two phases coexist. The boundary of the region of coexistence is shown as a black line.

Legendre transform with respect to all its arguments gives a function of the intensive variables T, P and μ alone which, by scaling, must be identically zero. This is the Gibbs–Duhamel relation. It determines the pressure P as a convex function of the remaining intensive coordinates, (T, μ):

$$PV = \mu N + TS - E. \tag{2.2}$$

The pressure is a convenient object to consider because all the terms on the rhs of Eq. (2.2) admit a simple representation in statistical mechanics. $-P$ is sometimes called the grand potential, e.g., [12]. Since the pressure is the Legendre transform of the internal energy with respect to S and N, the convexity of E then implies the convexity of the pressure with respect to T and μ.

Now, it is a consequence of the duality of the Legendre transform that if E has a linear segment of length ΔX, then its Legendre transform P has a corresponding jump in gradient with $\Delta(\nabla P) = \Delta X$. It follows that pure phases correspond to points where $P(T, \mu)$ has a unique tangent, while two phases coexist at those points (T, μ) where P has two (linearly independent) tangent planes. (Triple points are similarly defined.)

It is now a fact about convex functions that almost all points have a unique tangent while the set with multiple tangents has codimension 1 (in the sense of comparing Hausdorff dimensions). A geometric proof of this fact can be found in [6]. This gives a weak version of the Gibbs phase rule: If one considers the pressure P as a function of (T, μ), (or alternatively, chemical potential μ as a function of (P, t)), then pure phases are the typical sets while phases coexist on exceptional, i.e., small, sets.

2.3.2 Magnetic systems

At $T = 0$ the entropy term in Eq. (2.2) drops. For a system of non-interacting Fermions all single particle states below μ are occupied, while those above are empty. This says that for the single particle Hamiltonian H and area A the pressure is

$$P = \lim_{A \to \infty} \frac{1}{A} Tr \left((\mu - H)_+ \chi(A) \right);   \tag{2.3}$$

χ is the characteristic function of the area and $x_+ = x\, \theta(x)$ with θ a unit step function.

Let B denote the magnetic induction (i.e., the macroscopic average of the local magnetic field [11]). The Hamiltonian is a function of B and so is the pressure. The density ρ and the (specific) magnetization M are then given by [11]

$$\rho = \frac{\partial P}{\partial \mu}, \quad M = \frac{\partial P}{\partial B} .   \tag{2.4}$$

The Hall conductance is thermodynamically defined by

$$\sigma_H = \frac{\partial \rho}{\partial B} = \frac{\partial M}{\partial \mu} .   \tag{2.5}$$

It follows that, in the the wings of the butterflies where the Hall conductance is quantized, P is given by

$$P(\mu, B) = \sigma_g B(\mu - \mu_g) ,   \tag{2.6}$$

where g is a discrete wing label. The wings represent pure phases since P has a unique tangent in the gaps.

2.3.3 The order of the transitions

P, ρ and M are bi-linear in μ and B in the gaps. P and ρ are actually also continuous functions of μ on the spectrum. For rational flux this is a consequence of Floquet theory. For irrational flux this can be seen by a limiting argument.

At the same time, the Hall conductance, being integer-valued on a set of full measure, can not be extended to a continuous function.[2] (In fact, it is not even bounded.) The continuity of the first derivative and the discontinuity of the second derivatives makes the phase transitions in μ second order according to the Ehrenfest classification [3].

---

[2] The magnetization does not extend to a continuous function on the spectrum for rational fluxes [7].

2.3.4 Phases and their boundaries

In the colored Hofstadter butterflies, pure phases are open sets. The boundary of a given phase, say the red wing, is a curve; it is not a smooth curve as at rational values of B it has distinct tangents, but it is still a curve of Hausdorf dimension 1 [15]. This is reminiscent of the Gibbs phase rule. Note, however, that the notion of the boundary of a pure phase, and the notion of phase coexistence, are distinct. The phase with Hall conductance 1 meets the phase 0 at a single point, at the tip of the butterfly, not on a line, as one might expect by the Gibbs phase rule. This holds in general: The boundary of the phases i intersects the boundary of the phase j on a set of codimension 2, not 1 [15]. Moreover, any small disc that contains two distinct phases of the butterfly contain infinitely many other phases.

2.3.5 Magnetic domains and phase coexistence

Is the fractal phase diagram of the butterfly in conflict with basic thermodynamic principles?

The Gibbs phase rule one finds in classical thermodynamics [3] says that two phases meet on a smooth curve, which is clearly not the case for the butterfly. However, this strong version of the Gibbs rule involves assumptions of smoothness of free energies that may or may not hold. Convexity alone gives a weaker version of the Gibbs phase rule, which we briefly discussed in section 2.3.1, which allows for all kind of wild behaviors, and does not rule out fractal phase diagrams like the butterfly.[3]

More worrisome is the lack of convexity of the pressure, $P(\mu, B)$ which is manifest in the periodicity of Fig. 2.1 in B. This raises the question if this reflects a problem with the Hofstadter model. It does not. A little reflection shows that rather, it a consequence of choosing B, the magnetic induction, as the thermodynamic coordinate. In the remaining part of this section I shall explain why it is actually more natural to choose, for the independent thermodynamic variable, the magnetic field H and the difficulties in drawing the butterflies in the $\mu - H$ plane.

Imagine a two-dimensional system with finite width which is broken to domains. Assume that the magnetic field in each of the domains is perpendicular to the plane and is constant through the given domain. Since $\nabla \times H = 0$, the magnetic field H must be the same in adjacent domains. Hence the notion of constant magnetic field is constant H, while B will not be constant if the system breaks into domains. The problem with H constant is more difficult because it is B, not H, that enters in the Hamiltonian [11].

Given the colored butterfly as a function of B, what can one say about the colored butterfly as a function of H? Recall that B, H and the magnetization

---

[3] Instructive examples are given on p. 8 of [17]. I thank Aernout van Enter for pointing out this example to me and for a clarifying discussion on the Gibbs phase rule.

M are related by

$$B = H + 4\pi M \,. \tag{2.7}$$

Since M is a function of B, so is H. However, B may fail to be a (univalued) function of H. This is the case if $-4\pi\partial_B M \geq 1$; if the magnetic susceptibility is sufficiently negative. When this happens, the relation $H(B)$ can not be inverted to a function $B(H)$. Domains with different values of M and B may then form and coexist [11, 4].

The condition for coexistence is a stability condition: The system will pick a value of B, consistent with H, that will minimize the entropy. However, at $T = 0$ the entropy of a gas of Fermions vanishes, so the different solutions $B_j(H,\mu)$ all give the same entropy, zero. This suggests that all the B_j represent phases at coexistence.

There is no reason why this degeneracy will hold if T is not strictly 0. Then, for most values of H a distinguished solution of $B_0(H,\mu)$ will be picked. The simple scenario is that $B_0(H,\mu)$ will depend, for most H, continuously on H. In these intervals, the phases of the colored butterfly in (μ, H) will be a deformed version of the phases in (μ, B). However, since a value of $B_0(H,\mu)$ is picked by a minimization procedure, there is no guarantee for continuity and $B_0(H,\mu)$ will be, in general, a discontinuous function of its arguments. At the discontinuities, major qualitative changes in the diagram will take place and it is interesting to investigate the colored butterfly in the $\mu - H$ plane.

Another open problem in this context is to analyze the domain structure for coexisting phases. The quasi-periodic character of the electronic problem for irrational fluxes suggest that the domain structure could be rich and interesting as well, (e.g., a quasi-periodic domain structure for irrational fluxes).

2.4 Duality

We now turn to the duality relating the two diagrams.

2.4.1 Weak magnetic fields

Consider the "Bloch band" dispersion relation

$$\epsilon(k) = \cos(k \cdot a) + \cos(k \cdot b) \tag{2.8}$$

on the two-dimensional Brillouin zone. a, b are the unit lattice vectors. The Hamiltonian describing a weak external magnetic field is obtained by imposing the canonical commutation relation

$$[k \cdot a, k \cdot b] = ia \times b \cdot B = i\Phi. \tag{2.9}$$

This procedure is known as the Peierls substitution [16]. The model is known as the Harper model, after a student of Peierls. The spectrum, plotted in

Fig. 2.2, is a set of measure zero and so invisible in Fig. 2.1. Figs. 2.1 and 2.2 describe disjoint and complementary sets, whose union is the plane.

Although there is considerable interest in the Hofstadter model for its own sake (see e.g., [1] and references therein) its physical significance to the two-dimensional electron gas is limited. One reason is that the flux Φ, even for the strongest available magnetic fields, is tiny and only a horizontal sliver of the diagram in Fig. 2.1 near zero flux can be realized. Moreover, Φ of order 1 is presumably outside the region of weak field for which the model approximates the Schrödinger equation.

By gauge invariance, time-reversal and electron-hole symmetry, the pressure satisfies [7]

$$P(\mu, \Phi) = P(\mu, -\Phi) = P(\mu, \Phi + 1) = -\mu + P(-\mu, \Phi). \tag{2.10}$$

This gives Fig. 2.1 its symmetry.

2.4.2 Strong magnetic fields

A classical charged particle in a homogeneous magnetic field moves on a circle. The center of the circle is, classically,

$$c = x + \frac{v \times B}{B^2}; \tag{2.11}$$

c commutes with v, but the components of the center do not commute, rather they satisfy the canonical commutation relations

$$[c \cdot a^*, c \cdot b^*] = -i\frac{B \cdot a^* \times b^*}{B^2} = -i\frac{(2\pi)^2}{\Phi}. \tag{2.12}$$

(a^*, b^*) are dual vectors to (a, b).

If the wave function ψ belongs to a given Landau level, then the shifts $e^{ic \cdot \alpha}\psi$, for $\alpha \in \mathbb{R}^2$, span the spectral subspace of that level. This means that the Hamiltonian

$$\cos(c \cdot a^*) + \cos(c \cdot b^*), \tag{2.13}$$

acts within Landau levels. For large B it approximates the periodic potential $\cos(x \cdot a^*) + \cos(x \cdot b^*)$, which couples different Landau levels. This is seen from the fact that in a given Landau level $v = O(\sqrt{B})$, hence, by Eq. (2.11), $c \approx x$ for large B. The Hamiltonian in Eq. (2.12) is the same as that of Eq. (2.8), except that in the commutator $\Phi/2\pi$ of Eq. (2.9) is replaced by $2\pi/\Phi$ of Eq. (2.12).

Although the spectral problem of the two models is essentially the same, the phase diagrams are different. This is explained in the next subsection.

Unlike the tight-binding model which is mostly of academic interest, the model of a split Landau level can be realized in artificial superlattices that accommodate a unit of quantum flux at attainable fields.

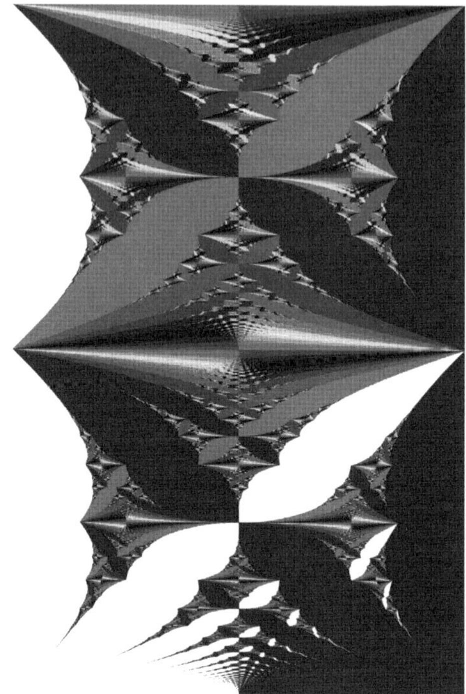

Fig. 2.4. Colored Hofstadter butterfly for Landau level split by a super-lattice periodic potential. The horizontal axis is the chemical potential; the vertical axis is the average number of unit cells associated with a unit of quantum flux. As the number increases by 1, the pattern repeats but with a different coloring code. (See color insert following p. 22.)

2.4.3 Thermodynamic duality

The pressure of a split Landau level, P_l, and split Bloch band P_b, for any temperature T, are related by [7]

$$P_l(T, \mu, \Phi/2\pi) = \frac{\Phi}{2\pi} P_b(T, \mu, 2\pi/\Phi) \,. \tag{2.14}$$

This is a duality transformation: It is symmetric under the interchange $b \longleftrightarrow l$. It implies that the thermodynamics of the split Bloch band determine the thermodynamics of a split Landau level and vice versa. The factor Φ on the right is the reason that P_l is not periodic, although P_b is.

A check on the factor $\Phi/2\pi$ comes by considering large μ. Then, the tight binding model has all sites occupied and the electron density is $\rho_b \to 1$. This implies $P_b \to \mu$. In contrast, a full Landau level has electron density that is proportional to the flux through unit area: $\rho_l \to \Phi/2\pi$ so $P_l \to \Phi\mu/2\pi$.

The magnetization and the Hall conductances of the two models are therefore related by:

$$m_l(\mu, T, 2\pi/\Phi) = -\frac{1}{2\pi}\, P_b(\mu, T, \Phi/2\pi) - \frac{\Phi}{2\pi}\, m_b(\mu, T, \Phi/2\pi);$$

$$\sigma_l(\mu, T, 2\pi/\Phi) = \frac{1}{2\pi}\, \rho_b(\mu, T, \Phi/2\pi) - \frac{\Phi}{2\pi}\, \sigma_b(\mu, T, \Phi/2\pi)\,. \qquad (2.15)$$

When μ is large, $\sigma_b = 0$, since a full band is an insulator. At the same time, a full Landau level has a unit of quantum conductance, $\sigma_l = 1/2\pi$, in agreement with the Eq. (2.15).

Acknowledgment: I thank A. van Enter and O. Gat for useful discussions and criticism. This work is supported by the Technion fund for promotion of research and EU grant HPRN-CT-2002-00277.

2.5 Appendix: Diophantine equation

Let me finally describe the algorithm of [18] for coloring the gaps in the butterfly Fig. 2.1. Suppose that the magnetic flux through a unit cell is $\frac{p}{q}$. For p and q relatively prime, define the conjugate pair (m, n) as the solutions of

$$pm - qn = 1\,. \qquad (2.16)$$

m is determined by this equation modulo q and n modulo p. The algorithm for solving Eq. (2.16) is the division algorithm of Euclid. (Standard computer packages for finding the greatest common divisor of p and q, yield also m and n such that $pm + qn = \gcd(p, q)$.) The Hall conductance k_j, associated with the j-th gap, in the tight binding case, is given by [18]

$$k_j = jm \bmod q, \quad |k_j| \le q/2\,. \qquad (2.17)$$

In the case of a split Landau band, Eq. (2.17) again determines k_j provided p and q are interchanged.

References

1. A. Avila and R. Krikorian, `http://xxx.lanl.gov/abs/math.DS/0306382`; J. Puig, Cantor spectrum for the almost Mathieu operator, mp-arc 03-145.
2. M.Ya. Azbel, *Sov. Phys. JETP* **19** (1964), 634–645.
3. H.B. Callen, *Thermodynamics and an Introduction to Thermostatics*, Wiley, 1985.
4. J.H. Condon, *Phys. Rev.* **145** (1966), 526.
5. R.H. Dalitz and Sir R. Peierls, *Selected Scientific Papers of Sir Rudolph Peierls*, World Scientific, 1997.

6. K. Falconer, *Fractal Geometry*, Wiley, 1990.

7. O. Gat and J. Avron, Magnetic fingerprints of fractal spectra, *New J. Phys.* **5** (2003), 44.1–44.8; http://arxiv.org/abs/cond-mat/0212647.

8. O. Gat and J. Avron, Semiclassical analysis and the magnetization of the Hofstadter model, *Phys. Rev. Lett.* **91** (2003), 186801; http://arxiv.org/abs/cond-mat/0306318.

9. D. Hofstadter, *Phys. Rev. B* **14** (1976), 2239–2249.

10. R. Israel, *Convexity in the Theory of Lattice Gases*, Princeton, 1979.

11. L.D. Landau and E.M. Lifshitz, Electrodynamics of continuous media, Butterworth-Heinemann, 1983.

12. E.M. Lifshitz and L.P. Pitaevskii *Statistical Physics*, Pergamon, 1980.

13. S.P. Novikov, JETP **52** (1980), 511.

14. D. Osadchy, M.Sc. thesis, Technion, 2001.

15. D. Osadchy and J. Avron, Hofstadter butterfly as quantum phase diagram, *J. Math. Phys.* **42** (2001), 5665–5671; http://arxiv.org/abs/math-ph/0101019.

16. G. Panati, H. Spohn and S. Teufel, *Phys. Rev. Lett.* **88** (2002), 250405; and http://arxiv.org/abs/math-ph/0212041.

17. A.W. Roberts and D.E. Varberg, *Convex Functions*, Academic Press, 1973.

18. D.J. Thouless, M. Kohmoto, M.P. Nightingale, and M. den Nijs, *Phys. Rev. Lett.* **49** (1982), 405–408.

3

Semiclassical Normal Forms

D. Bambusi

Dipartimento di Matematica
Via Saldini 50
I-20133 Milano, Italy

Summary. Given a classical Hamiltonian function having an absolute minimum, we consider the problem of describing in the semiclassical limit the lowest part of the spectrum of the corresponding quantum operator. To this end we present an extension of the classical Birkhoff normal form to the semiclassical context and we use it to deduce spectral information on the quantum Hamiltonian. The properties of the spectrum turn out to be strongly dependent on the resonance relations fulfilled by the frequencies of small oscillations of the classical system. Here we concentrate on two opposite cases, namely the completely nonresonant and the completely resonant one and describe the spectrum of the Hamiltonian in these cases.

3.1 Introduction

Consider a Hamiltonian system with Hamiltonian function

$$H(\xi, x) := H_0 + V(x) , \quad H_0(\xi, x) := \sum_{i=1}^{n} \frac{\xi_i^2 + \omega_i^2 x_i^2}{2} , \tag{3.1}$$

where $V(x)$ is a C^∞ function having a zero of at least third order at the origin, and ω_l are the frequencies of small oscillation.

In the classical case it is well known that there exists a canonical transformation putting the system in Birkhoff normal form up to small remainder. Such a normal form can be used to describe the small amplitude solutions of the system. A corresponding quantum procedure has been recently introduced in [BV90, GP87, Bam95, Sjo92, BGP99, BCT03] (see also [Pop00a, Pop00b]). Accordingly one can construct a unitary transformation putting the quantum system in normal form. Such a normal form can be used to deduce spectral information on the Hamiltonian operator. Moreover such information remains valid in the semiclassical limit.

In this article we will present the results of [Bam95, BGP99, BCT03] in a unified way. First we will introduce at a purely formal level the classical and the quantum algorithm that put the system in normal form. Then we will show how to obtain a unified treatment. Precisely we will construct a unitary transformation that transforms the quantum Hamiltonian operator of the system into a normalized one. We will show that, when $\hbar \to 0$ the symbol of the normalized operator reduces to the classical normal form of the Hamiltonian (3.1).

Subsequently we will make the theory rigorous by adding estimates of the remainder. Such estimates are uniform in the semiclassical limit. This will be done in three steps. Each step will deal with a case that is more realistic, but technically more difficult, than the previous one. The result is that the low-lying eigenvalues of the normal form operator and those of the original Hamiltonian operator are close to each other.

In order to deduce precise spectral information one has to study the spectrum of the normal form: We will concentrate on two particular cases, namely the completely nonresonant and the completely resonant case.

In the nonresonant case the normal form has the structure

$$\hat{H}_0 + \hat{Z}_Q \tag{3.2}$$

where

$$\hat{H}_0 := -\frac{\hbar^2}{2}\Delta + \sum_j \omega_j^2 \frac{x_j^2}{2} \tag{3.3}$$

is the quantization of the harmonic oscillators, and $Z_Q = Z_Q(\hat{I}_1, \ldots, \hat{I}_n)$ is a function of the operators

$$\hat{I}_j := \frac{1}{\omega_j}\left(-\frac{\hbar^2}{2}\frac{\partial^2}{\partial x_j^2} + \omega_j^2 \frac{x_j^2}{2}\right).$$

The eigenvalues of (3.2) are given by

$$\hbar\omega \cdot \left(k + \frac{1}{2}\right) + Z_Q\left(\hbar\left(k_1 + \frac{1}{2}\right), \ldots, \hbar\left(k_n + \frac{1}{2}\right)\right), \quad k \in \mathbb{N}^n. \tag{3.4}$$

This will give a quantization formula for the eigenvalues of the original operator.

We come to the completely resonant case. Here the frequencies are integer multiples of a single one, say ν. The normal form has again the structure (3.2), but the only available information on the operator $\hat{Z}_Q$ is that it commutes with $\hat{H}_0$.

To describe our procedure and results, we remark that in the linear approximations the spectrum of the system is given by $(n + 1/2)\nu\hbar$, $n \in \mathbb{N}$. Moreover the eigenvalues have a multiplicity that grows with n. The first effect of the nonlinearity is expected to be a splitting of the degeneracy and a transformation of a single eigenvalue into a small band of eigenvalues close to each other. We will show that the width of the band can be computed in the semiclassical limit ($\hbar \to 0$ with $n\hbar \to E$) using the normal form of the system. The idea is that the restriction of $\hat{Z}_Q$ to an eigenspace of $\hat{H}_0$ is a Toeplitz operator whose principal symbol is the restriction of the classical normal form to a surface of constant harmonic energy. So, using standard properties of Toeplitz operators one has that the extrema of such a restricted normal form are essentially the lowest and the highest eigenvalues of the above operator.

Related results were obtained by Zhilinski and coworkers [Zhi89, VSZB01] and by San Vũ Ngọc in his thesis [Vun98].

3.2 Formal theory

3.2.1 Classical Lie transform

We begin by recalling the classical Lie transform and its use to put systems in normal form.

To make the discussion clearer we introduce an artificial parameter ϵ, and, instead of the Hamiltonian (3.1) we consider, on $\mathbb{R}^{2n}$, the system

$$H_\epsilon := H_0 + \epsilon V . \tag{3.5}$$

We fix a large M and look for a canonical transformation putting (3.5) in normal form up to terms of order ϵ^M.

Let χ be a smooth function defined on $\mathbb{R}^{2n}$, and denote by

$$X_\chi \equiv \left(-\frac{\partial \chi}{\partial x}, \frac{\partial \chi}{\partial \xi} \right) , \quad (\xi, x) \in \mathbb{R}^{2n}$$

the corresponding Hamiltonian vector field. Consider the Hamiltonian equations of χ, namely

$$\dot{z} = X_\chi(z) , \quad z \equiv (\xi, x) \tag{3.6}$$

and let Φ_t be the corresponding flow.

Definition 1. *The map $\Phi := \Phi_\epsilon \equiv \Phi_t\big|_{t=\epsilon}$ will be called a Lie transform generated by χ.* $\qquad\square$

Remark 1. Φ is a canonical transformation of the form

$$\Phi(z) = z + \epsilon X_\chi(z) + O(\epsilon^2) .$$

Given a function g one can consider the transformed function $g \circ \Phi$. Using the equality

$$\frac{d}{dt}[g \circ \Phi_t] = \{\chi; g\} \circ \Phi_t,$$

it is easy to see that

$$g \circ \Phi = \sum_{l=0}^{\infty} \epsilon^l g_l \, ,$$

where the g_l's are defined by

$$g_l = \frac{1}{l} \{ \chi; g_{l-1} \} \, , \quad l \geq 1 \, ; \quad g_0 = g \, . \tag{3.7}$$

We want to choose a function χ such that $H_\epsilon \circ \Phi$ is in normal form up to order ϵ^2. By explicit computation one has

$$H_\epsilon \circ \Phi = H_0 + \epsilon [V + \{ \chi; H_0 \}] + O(\epsilon^2),$$

so χ has to fulfill the so-called homological equation:

$$V + \{ \chi; H_0 \} = Z \tag{3.8}$$

where Z Poisson commutes with H_0, i.e.,

$$\{ H_0; Z \} \equiv 0 \, .$$

It is well known that, at least formally equation (3.8) is always solvable, however the properties of Z strongly depend on the resonance relations fulfilled by the frequencies. In particular, if the frequencies are nonresonant, then Z turns out to be a function of the actions $I_j = (\xi_j^2 + \omega_j^2 x_j^2)/2\omega_j$ only. On the contrary, in the completely resonant case the only available information on Z is that it Poisson commutes with H_0. So one can use the function χ solving (3.8) to generate the Lie transform and to put the Hamiltonian in the form

$$H^{(1)} = H_0 + \epsilon Z + \epsilon^2 V^{(1)} \tag{3.9}$$

with a suitable $V^{(1)}$. Then one can iterate the construction and after M steps one gets a Hamiltonian of the form

$$H^{(M)} = H_0 + \epsilon Z^{(M)} + \epsilon^{M+1} V^{(M)} \tag{3.10}$$

where the nonnormalized part has been pushed to an arbitrarily high order.

3.2.2 Quantum Lie transform

Let W be a self-adjoint operator on $L^2(\mathbb{R}^n)$, and consider the corresponding unitary operator X_t defined by:

$$\dot{X}_t = \frac{1}{i\hbar} X_t W \, .$$

Definition 2. *The unitary operator* $X := X_\epsilon \equiv X_t\big|_{t=\epsilon}$ *will be called the quantum Lie transform generated by* W.

Given a linear operator G, consider the transformed operator XGX^{-1}; exploiting the equation

$$\frac{d}{dt}(X_t G X_t^{-1}) = X_t \frac{i}{\hbar}\,[W;G]\,X_t^{-1}\ ,$$

one has that the Taylor expansion of XGX^{-1} in ϵ is given by

$$XGX^{-1} = \sum_{l=0}^{\infty} \epsilon^l G_l\ ,$$

where

$$G_l = \frac{1}{l}\frac{1}{i\hbar}[W, G_{l-1}]\ ,\quad l \geq 1\ ;\quad G_0 = G\ . \tag{3.11}$$

Again one can use the quantum Lie transform to put a Hamiltonian of the form

$$\hat{H}_\epsilon = \hat{H}_0 + \epsilon\hat{V} \tag{3.12}$$

in normal form. Proceeding as in the classical case one has

$$X\hat{H}_\epsilon X^{-1} = H_0 + \epsilon\left(\hat{V} + \frac{i}{\hbar}\left[W;\hat{H}_0\right]\right) + O(\epsilon^2)$$

and therefore, if one is able to solve the quantum homological equation, namely to determine W and $\hat{Z}_Q$ such that

$$\hat{V} + \frac{i}{\hbar}\left[W;\hat{H}_0\right] = \hat{Z}_Q\ ,\quad \frac{i}{\hbar}\left[\hat{H}_0;\hat{Z}_Q\right] = 0, \tag{3.13}$$

then one has that the system is in normal form up to terms of order ϵ^2. Iterating the construction, one has that the system can be put in normal form up to a remainder of order ϵ^M.

It is clear that formally the two procedures are very similar. The true correspondence can be established by the use of Weyl quantization.

3.2.3 Semiclassical Lie transform

Given a function g on the classical phase space, one can define the corresponding quantum operator $Op^w(g)$ by

$$(Op^w(g)\psi)(x) = (2\pi\hbar)^{-n} \int_{\mathbb{R}^n \times \mathbb{R}^n} g((x+y)/2, \xi)e^{i\langle x-y,\xi\rangle/\hbar}\psi(y)\,dy\,d\xi\ . \tag{3.14}$$

It is possible to show that, under suitable assumptions, $Op^w(g)$ is well defined and fulfills the standard properties of the quantization by the symmetrization rule (for a systematic description of this quantization procedure see, e.g.,

28   D. Bambusi

[BS91, Ro87]). Due to formula (3.14) one has a correspondence between a suitable class of functions and self-adjoint operators in L^2. The function corresponding to a given operator (when it exists) is called *the symbol* of the operator. The main point is that, given two functions g_1 and g_2, there exists a unique function whose Weyl quantization is $\frac{i}{\hbar}[Op^w(g_1); Op^w(g_2)]$. This function is the so-called Moyal bracket of g_1 and g_2, and is usually denoted by $\{g_1; g_2\}_M$. By formally expanding the Moyal bracket in powers of $\hbar$, one gets [BS91]

$$\{g_1; g_2\}_M = \{g_1; g_2\} + O(\hbar^2) . \tag{3.15}$$

Moreover it can be shown that if one of the functions g_1 or g_2 is a polynomial of degree at most 2, then

$$\{g_1; g_2\}_M = \{g_1; g_2\}$$

so that, in particular $\{H_0; g\}_M = \{H_0; g\}$ for any function g. It is possible to reformulate the quantum normalization procedure of the previous section in terms of Moyal brackets. Indeed, given two functions χ and g, define the sequence $\{\tilde{g}_l\}_{l \geq 0}$ by

$$\tilde{g}_0 := g , \quad \tilde{g}_l := \frac{1}{l}\{\chi; \tilde{g}_{l-1}\}_M \tag{3.16}$$

and the function

$$\tilde{g} := \sum_{l=0}^{\infty} \epsilon^l \tilde{g}_l . \tag{3.17}$$

Then (formally) $\tilde{g} = g \circ \Phi + O(\hbar^2)$, where Φ is the classical Lie transform generated by χ. Moreover, comparing with the formulae of the previous section one has

$$X Op^w(g) X^{-1} = Op^w(\tilde{g}) \tag{3.18}$$

where X is the quantum Lie transform generated by $Op^w(\chi)$. Repeating the construction of the previous section in terms of symbols of the operators, one has the following

Proposition 1. *For any M there exists a formal unitary transformation conjugating the Hamiltonian operator $Op^w(H_\epsilon)$ (with H_ϵ given by (3.5)) to an operator with symbol of the form*

$$H_0 + \epsilon Z_Q^{(M)} + \epsilon^{M+1} V^{(M)}$$

where $Z_Q^{(M)} = Z_Q^{(M)}(x, \xi, \hbar)$ fulfills

$$\left\{Z_Q^{(M)}, H_0\right\} = \left\{Z_Q^{(M)}, H_0\right\}_M = 0,$$

and moreover $Z_Q^{(M)}(x, \xi, 0)$ coincides with the Birkhoff normal form $Z^{(M)}$ of the classical Hamiltonian system (3.5).

3.3 Rigorous theory

3.3.1 The case of analytic fast decreasing perturbations

Up to now the theory has been developed at a heuristic level, indeed we paid no attention to the problem of convergence of series or to that of estimating the remainder $V^{(M)}$. In this subsection we will give some rigorous results on the spectrum in the technically simplest situation. The main point consists in introducing a class of functions which behaves well under the operations involved in the construction of the normal form, namely quantum Lie transform and solution of the homological equation. To obtain the wanted class of functions, we begin by defining an analytic action Ψ of $\mathbb{T}^n$ into $\mathbb{R}^{2n}$ through the flow of H_0:

$$\Psi \;:\; \mathbb{T}^n \times \mathbb{R}^{2n} \to \mathbb{R}^{2n},$$
$$(\varphi, (\xi, x)) \mapsto (\xi', x') = \Psi_\varphi(\xi, x),$$
$$x'_k := \frac{\xi_k}{\omega_k} \sin \varphi_k + x_k \cos \varphi_k,$$
$$\xi'_k := \xi_k \cos \varphi_k - \omega_k x_k \sin \varphi_k \,.$$

Remark that, denoting $z := (\xi, x)$, the flow $\phi^t(z)$ of X_{H_0} is $\phi^t(z) = \Psi_{\omega t}(z)$. Define the angular Fourier coefficient of order k of a function g by

$$(g)_k(z) := \frac{1}{(2\pi)^n} \int_{\mathbb{T}^n} g(\Psi_\varphi(z)) e^{-i\langle k, \varphi \rangle} \, d\varphi, \quad k \in \mathbb{Z}^n \,.$$

Remark 2. If $g \in \mathbb{C}^1$, then

$$g(\Psi_\varphi(z)) = \sum_{k \in \mathbb{Z}^n} (g)_k(z) e^{i\langle k, \varphi \rangle} \implies g(z) = \sum_{k \in \mathbb{Z}^n} (g)_k(z) \,.$$

Given $\rho > 0, \sigma > 0$, denote by $\widehat{(g)_k}(s)$ the space of Fourier transforms of $(g)_k(z)$ and define

$$\|g\|_{\rho,\sigma} := \sum_{k \in \mathbb{Z}^n} e^{\rho|k|} \int_{\mathbb{R}^{2n}} |\widehat{(g)_k}(s)| e^{\sigma|s|} \, ds \,. \tag{3.19}$$

Definition 3. *The space of the analytic functions such that the norm* (3.19) *is finite will be denoted by* $\mathcal{A}_{\rho,\sigma}$.

Remark 3. A function $f(\xi, x)$ of class $\mathcal{A}_{\rho,\sigma}$ in particular is integrable over the whole of $\mathbb{R}^{2n}$.

By standard results on Weyl quantization one has

$$\|Op^w(g)\|_{L^2 \to L^2} \leq \|g\|_{\rho,\sigma} \,. \tag{3.20}$$

The class $\mathcal{A}_{\rho,\sigma}$ behaves well under the quantum Lie transform. Indeed one has

Proposition 2. [BGP99] *Let $g \in \mathcal{A}_{\rho,\sigma}$ and $\chi \in \mathcal{A}_{\rho,\sigma}$ be two smooth symbols, fix $\delta < \sigma$ and assume that*

$$\epsilon \|\chi\|_{\rho,\sigma} \le \delta^2;$$

then the series $\tilde{g}_l$ defined by (3.16), (3.17) converges in the norm of $\mathcal{A}_{\rho,\sigma-\delta}$, and one has

$$XOp^w(g)X^{-1} = \sum_{l \ge 0} Op^w(\tilde{g}_l)\epsilon^l$$

where X is the quantum Lie transform generated by $Op^w(\chi)$ and the series on the r.h.s. converges also in the operator norm.

In order to solve the homological equation we have to make some assumptions on the frequencies.

H1)   There exist $\gamma > 0$ and $\tau \in \mathbb{R}$ such that, for any $k \in \mathbb{Z}^n$, one has

$$\text{either} \quad \omega \cdot k = 0 \quad \text{or} \quad |\omega \cdot k| \ge \frac{\gamma}{|k|^\tau}. \tag{3.21}$$

Proposition 3. [BGP99] *Assume H1, let $g \in \mathcal{A}_{\rho,\sigma}$. Then the quantum homological equation*

$$\{H_0, \chi\}_M + Z = g \tag{3.22}$$

admits solutions $\chi, Z \in \mathcal{A}_{\rho-d,\sigma}$, where $d < \rho$ is a positive parameter. Moreover one has $\{H_0, Z\} = 0$ and:

$$\|Z\|_{\rho,\sigma} \le \|g\|_{\rho,\sigma}; \quad \|\chi\|_{\rho-d,\sigma} \le c_\psi \frac{\|g\|_{\rho,\sigma}}{d^\tau}; \qquad c_\psi := \left(\frac{\tau}{e}\right)^\tau \frac{1}{\gamma}.$$

Combining the above propositions and the iterative algorithm introduced in the previous section one gets

Theorem 1. *Consider the Hamiltonian operator (3.12), assume $V \in \mathcal{A}_{\rho,\sigma}$, fix $\delta < \min\{\rho, \sigma\}$, and assume also that H1 holds. There exist constants C_1, C_2, such that, for any integer $M > 0$, the following holds true: if*

$$\frac{\epsilon \|V\|_{\rho,\sigma}}{\delta^{2+\tau}} < 1/C_1 M^{2+\tau}, \tag{3.23}$$

then there exists a unitary transformation conjugating the system (3.12) to

$$Op^w(H_0) + \epsilon Op^w(Z_Q^{(M)}) + R_M$$

where $Z_Q^{(M)}(z, \epsilon, \hbar) \in \mathcal{A}_{\rho-\delta,\sigma-\delta}$ is in normal form. $Z_Q^{(M)}(z, \epsilon, 0)$ coincides with the classical Birkhoff normal form of $H_0 + \epsilon V$ computed up to order M. Moreover, one has

$$\|R_M\|_{L^2 \to L^2} \le \frac{C_1}{\delta^{2+\tau}} \left(C_2 M^{2+\tau} \epsilon \|V\|_{\rho,\sigma}\right)^{M+1}. \tag{3.24}$$

Remark 4. Choosing $M = M_{\mathrm{opt}} \sim \epsilon^{-1/(2+\tau)}$, one gets an exponentially small estimate of the remainder:

$$\|R_{M_{\mathrm{opt}}}\|_{L^2 \to L^2} \leq C_1 \exp\left(-\frac{C_2}{\epsilon^{1/(2+\tau)}}\right) \tag{3.25}$$

with redefined constants C_1, C_2. In the following we will concentrate on this kind of exponential estimates.

Remark 5. In the nonresonant case where $\omega \cdot k \neq 0$ for $k \neq 0$, one has

$$Z_Q^{(M)} = Z_Q^{(M)}(I_1, \ldots, I_n, \epsilon, \hbar) ,$$

i.e., the function $Z_Q^{(M)}$ depends only on the actions.

3.3.2 The case of compactly supported perturbations

The applicability of the above theorem is not clear, since the definition of the class $\mathcal{A}_{\rho,\sigma}$ is quite implicit. In the present subsection we will extend the result to the case of C^∞ compactly supported perturbations of class Gevrey (precisely fulfilling the assumption HG below); in the next subsection we will show how to deal with the general cases. So we assume that $V = V(x, \xi)$ has compact support and is of class Gevrey, namely that

(HG) there exist $\alpha > 0, K > 0$ and $1 < \ell < \infty$ such that $\forall j = (j_1, \ldots, j_{2n})$,

$$\sup_{z \in \mathbb{R}^{2n}} \left|\frac{\partial^{|j|}}{\partial z^j} V(z)\right| \leq K \alpha^{-|j|}(|j|!)^\ell , \quad z = (\xi, x) . \tag{3.26}$$

Theorem 2. *Consider the operator (3.12), assume H1 and HG. Then there exist constants $C_1, C_2, b > 0$ such that the following holds true: if ϵ is small enough, then there exists a unitary transformation conjugating the system (3.12) to*

$$Op^w(H_0) + \epsilon Op^w(Z_Q) + R_{opt}$$

where $Z_Q(z, \epsilon, \hbar)$ is in normal form. $Z_Q(z, \epsilon, 0)$ coincides with the classical Birkhoff normal form of $H_0 + \epsilon V$ computed up to a suitable ϵ dependent order. Moreover, one has

$$\|R_{opt}\|_{L^2 \to L^2} \leq C_1 \epsilon \exp(-C_2/\epsilon^b) . \tag{3.27}$$

Idea of the proof. For $\Lambda > 0$ and $K > 0$, approximate V by

$$V^{\Lambda,K}(z) := \sum_{|k| \leq K} \int_{|s| < \Lambda} \widehat{(V)_k}(s) e^{i\langle s, z\rangle} \, ds \tag{3.28}$$

and apply Theorem 1 (with exponential estimate of the remainder) to $H_0 + V^{\Lambda,K}$, keeping track of the dependence of all the constants on Λ and K. Then choose Λ and K in such a way that the estimate (3.25) of the remainder is of the same order of magnitude as the error due to the approximation of V by $V^{\Lambda,K}$. This gives the result. For details see [BGP99].

3.3.3 The case of an unbounded potential

Using microlocal analysis it is possible to reduce a quite general case to the previous one. The idea is to consider a quantum Hamiltonian of the form

$$Op^w(H) \ , \quad H := H_0 + V \tag{3.29}$$

with a possibly unbounded $V = V(x)$ having a zero of third order at the origin and then to substitute the potential V with a different function, say V_ϵ, with support contained in a ball of radius 2ϵ in the phase space and coinciding with the original one in a ball of radius ϵ. One expects the lowest parts of the spectra of

$$Op^w(H) \equiv Op^w(H_0 + V) \quad \text{and} \quad Op^w(H_0 + V_\epsilon)$$

to be close to each other. This is exactly the content of the forthcoming results. The precise relation between the spectra is captured by the following definition. Let $H_1(\kappa), H_2(\kappa)$ be two families of self-adjoint operators depending on the parameters $\kappa \in \mathcal{D} \subset \mathbb{R}^k$, let $J_\kappa \subset \mathbb{R}$ be a family of bounded interval, and set $\mathrm{Spec}_{J_\kappa}(H_{1,2}) := \mathrm{Spec}(H_{1,2}) \cap J_\kappa$.

Definition 4. *Let $g : \mathcal{D} \to \mathbb{R}$ be a real function vanishing at $\bar{\kappa}$. We say that* $\mathrm{Spec}_{J_\kappa}(H_1) = \mathrm{Spec}_{J_\kappa}(H_2) \pmod{g}$ *as $\kappa \to \bar{\kappa}$ if*

$$H_1(\kappa)\phi_1(\kappa) = \lambda_1(\kappa)\phi_1(\kappa) \ , \quad \phi_1 \neq 0 \ , \quad \lambda_1 \in J_\kappa \tag{3.30}$$

imply

$$\|[H_2(\kappa) - \lambda_1(\kappa)]\phi_1(\kappa)\| \leq g(\kappa) \tag{3.31}$$

and, conversely.

Remark 6. If $\mathrm{Spec}_{J_\kappa}(H_1) = \mathrm{Spec}_{J_\kappa}(H_2) \mod g$, then for any $\lambda_1 \in \mathrm{Spec}_{J_\kappa}(H_1)$ there exists $\lambda_2 \in \mathrm{Spec}_{J_\kappa}(H_2)$ such that $\lambda_1(\kappa) = \lambda_2(\kappa) + Cg(\kappa))$ and conversely with a constant C uniform on J_κ.

In order to use microlocal analysis we need the following assumptions:

H2) $H \to +\infty$, $|x| \to \infty$; $H(x, \xi) > 0$ for $(x, \xi) \neq 0$.

H3) There exist $\nu \in \mathbb{R}$ $C_j > 0, \alpha > 0, K(U) > 0$ and $1 < \ell < \infty$ such that, $\forall j = (j_1, \dots, j_n)$, $|j| := |j_1| + \cdots + |j_n|$, and for any bounded $U \subset \mathbb{R}^n$:

$$\sup_{x \in \mathbb{R}^n} \left| (1 + |x|^2)^{\nu - |j|/2} \partial_{x_j}^{|j|} V(x) \right| \leq C_j,$$

$$\sup_{x \in U} \left| \partial_{x_j}^{|j|} V(x) \right| \leq K(U)\alpha^{-|j|}(|j|!)^\ell. \tag{3.32}$$

We are now ready for the application of the results of the previous sections. This is obtained in four steps.

1) We rescale the variables using the unitary transformation

$$U_\epsilon : L^2(\mathbb{R}^n) \ni \psi(x) \mapsto (U_\epsilon\psi)(x) := \epsilon^{n/2}\psi(\epsilon x) \in L^2(\mathbb{R}^n) \,. \qquad (3.33)$$

The image of $Op^w(H)$ under U_ϵ is the Weyl quantization of

$$\epsilon^2 H_\epsilon := \epsilon^2(H_0 + \epsilon\tilde{V}) \,, \quad \tilde{V}(x) := \epsilon^{-3}V(\epsilon x) \,, \qquad (3.34)$$

but a Weyl quantization where $\hbar$ is substituted by $\hbar' := \hbar/\epsilon^2$.

2) We make a cutoff of H_ϵ, namely, fix R and consider a Gevrey function t such that $t(s) \equiv 1$ for $|s| \leq R$, $t(s) \equiv 0$ for $|s| \geq 2R$, and define

$$a_{\epsilon,R}(x, \xi) := \tilde{V}(x)t(|x|)t(|\xi|) \,. \qquad (3.35)$$

3) We compare the spectrum of the Hamiltonian $Op^w(H_\epsilon)$ with the spectrum of $Op^w(H^t)$ where

$$H^t := H_0 + \epsilon a_{\epsilon,R} \,. \qquad (3.36)$$

This is done by the following

Proposition 4. [BGP99] *Assume H2,H3, let $J \subset \mathbb{R}_+$ be bounded and fixed. Then there exist $R > 0$ and $\epsilon_* > 0$ independent of $\epsilon, \hbar'$, such that, provided $0 \leq \epsilon \leq \epsilon_*$, one has*

$$\mathrm{Spec}_J(H_\epsilon) = \mathrm{Spec}_J(H^t) \ (\mathrm{mod}\ e^{-c/\hbar'^{1/\ell}}) \,.$$

4) Rescale back the variables, namely apply the transformation U_ϵ^{-1} to H^t. Defining $V_\epsilon := V(x)t(|x|/\epsilon)t(|\xi|/\epsilon)$, one has

$$\epsilon^2 U_\epsilon H^t U_\epsilon^{-1} = Op^w(H_0) + Op^w(V_\epsilon) \,.$$

It follows that

$$\mathrm{Spec}_{\epsilon^2 J}(Op^w(H_0 + V)) = \mathrm{Spec}_{\epsilon^2 J}(Op^w(H_0 + V_\epsilon)) \ (\mathrm{mod}\ e^{-c(\epsilon^2/\hbar)^{1/\ell}})$$

5) Now apply Theorem 2 to $H_0 + V_\epsilon$. One has

Theorem 3. *Assume H1,H2,H3. Then there exist $b > 0, h_* > 0, \epsilon_* > 0, A > 0, B > 0$ (independent of $\hbar$ and ϵ) and a smooth function*

$$Z(\xi, x; \hbar, \epsilon) : \mathbb{R}^{2n} \times [0, h_*] \times [0, \epsilon_*] \to \mathbb{R}$$

which commutes with H_0, such that, $\forall \epsilon \leq \epsilon_*, \hbar \leq \hbar_*\epsilon^2$ one has

$$\mathrm{Spec}_{\epsilon^2 J}(Op^w(H_0 + V)) = \mathrm{Spec}_{\epsilon^2 J}(Op^w(H_0 + Z))$$
$$\mathrm{mod}\ \left(e^{-c(\epsilon^2/\hbar)^{1/\ell}} + \epsilon^3 e^{-A/\epsilon^b}\right) \,.$$

Moreover Z admits a full asymptotic expansion in $\hbar$, has support in a ball of radius 2ϵ and $Z(\xi, x, 0, \epsilon)$ coincides in a ball of radius ϵ with the classical Birkhoff normal form truncated at an ϵ dependent order.

In the nonresonant case one thus gets a complete description of the spectrum of the Hamiltonian operator.

Corollary 1. *[BGP99] With the same assumptions and notation of Theorem 3, assume also that the frequencies are nonresonant, then the eigenvalues of $Op^w(H)$ in $[0, \epsilon^2]$ have the representation*

$$\langle (k+1/2), \omega \rangle \hbar + Z\left((k+1/2)\hbar, \hbar, \epsilon\right) + O\left(e^{-(B\epsilon^2/\hbar)^{1/\ell}}\right) + O\left(\epsilon^3 e^{-A/\epsilon^b}\right) \tag{3.37}$$

with $k \in \mathbb{N}^n$.

3.4 The Resonant Case

In this section we study the case where the linear frequencies are completely resonant, i.e., there exists $\nu \in \mathbb{R}_+$ and integers l_j such that $\omega_j = \nu l_j$. In this case it is nontrivial to study the spectrum of $H_0 + Z$. A first qualitative change in the spectrum is induced by the first nonlinear term of Z. Assume for simplicity that

$$Z = N^t + R$$

with N^t coinciding in a ball of radius ϵ with a polynomial of order 4 and R having a zero of order 5 at the origin. We will study the spectrum of the Weyl quantization of

$$H_0 + N^t. \tag{3.38}$$

Remark that the spectrum of $Op^w(H_0)$ is given by

$$\mathrm{Sp}(Op^w(H_0)) = \{\hbar\nu(\langle k, l \rangle + \frac{1}{2}|l|),\ k \in \mathbb{N}^n\} \subset \nu\hbar(\mathbb{N} + \frac{1}{2}|l|)\ .$$

Fix $E \in \mathrm{Sp}(Op^w(H_0))$, $E < \epsilon^2$ and remark that $Op^w(N^t)$ restricts to a well-defined operator N_E^t on the eigenspace corresponding to E. The remarkable fact is that such an operator can be realized as a Toeplitz operator whose principal symbol coincides with the restriction of N^t to the level surface $H_0 = E$. To come to this result we recall some facts about classical and quantum reduction.

3.4.1 Bargmann transform

We define the Bargmann transform $L_\hbar$ acting on $L^2(\mathbb{R}^n)$ by

$$(L_\hbar \psi)(z) := \left(\frac{1}{4\pi\hbar}\right)^{n/4} \int_{\mathbb{R}^n} e^{\frac{1}{2\hbar}(z^2 + 2\sqrt{2}x\cdot z - x^2)} \psi(x)\, dx\ .$$

It can be shown that $L_\hbar$ is an isomorphism between $L^2(\mathbb{R}^n)$ and the Bargmann space $\mathcal{B}_\hbar$ of entire holomorphic functions f such that the following integral is finite,

$$\int_{\mathbb{C}^n} |f|^2 e^{-|z|^2/\hbar} dz d\bar{z} \ . \tag{3.39}$$

Remark 7. The Bargmann space is particularly adapted to harmonic oscillators. Indeed one has

$$L_\hbar Op^w(H_0)L_\hbar^{-1} = \nu\hbar \sum_j l_j \left(z_j \frac{\partial}{\partial z_j} + \frac{1}{2} \right) \tag{3.40}$$

so that it is immediate to see that, having fixed a positive integer k, the eigenspace $\mathcal{E}_k$ of $Op^w(H_0)$ corresponding to $E_k = \nu\hbar(k + |l|/2)$ is

$$\mathcal{E}_k := \text{Span}\,\{z^\alpha\}_{\alpha\cdot l=k} \ . \tag{3.41}$$

For any Weyl symbol $P(\xi, x)$, one has that $L_\hbar Op^w(P)L_\hbar^{-1}$ is the Weyl quantization of

$$P_b(z, \bar{z}) := P\left(\frac{z + \bar{z}}{\sqrt{2}}, \frac{z - \bar{z}}{i\sqrt{2}} \right) \ .$$

The main point for our developments is that such an operator is a Toeplitz operator.

3.4.2 Geometric quantization and Toeplitz operators

Consider a $2n$-dimensional symplectic manifold $\mathcal{P}$ and assume that it admits a compatible complex structure. Assume also that it is endowed with a pre-quantization line bundle $L \to \mathcal{P}$, i.e., a Hermitian line bundle fulfilling some compatibility conditions. To fix ideas one can think of $\mathcal{P}$ as $\mathbb{C}^n$ endowed by the coordinates $z, \bar{z}$ and of L as the trivial bundle $\mathbb{C} \times \mathbb{C}^n \to \mathbb{C}^n$. For any k consider the Hilbert space of the *holomorphic* sections of $L^{\otimes k}$,

$$\mathcal{H}_k := \left\{ \Psi : \mathcal{P} \to L^{\otimes k} \right\} \ .$$

Here $1/k$ plays the role of $\hbar$. We assume that $\mathcal{H}_k$ is a closed subspace of the space $L^2(\mathcal{P}, L^{\otimes k})$ of the L^2 sections of $\mathcal{P}^{\otimes k}$, then there exists an orthogonal projector $\Pi_k : \mathcal{H}_k \to L^2(\mathcal{P}, L^{\otimes k})$.

Definition 5. *Given a function $f \in C^\infty(\mathcal{P})$, consider the multiplication operator*

$$M_f : \mathcal{H}_k \to L^2(\mathcal{P}, L^{\otimes k}),$$
$$\psi \mapsto f\psi$$

where the Toeplitz operator T_f with symbol f is defined by $T_f := \Pi_k M_f$.

In the example quoted above where $\mathcal{P} \equiv \mathbb{C}^n$, one can prove that one has $\mathcal{H}_k \equiv \mathcal{B}_{k-1}$, and, given a Weyl symbol $P(\xi)$, one has

$$L_{k-1}Op^w(P)L_{k-1}^{-1} = T_F \tag{3.42}$$

where

$$F(z, \bar{z}, k) = \left(e^{\frac{1}{k}\Delta}P_b\right)(z, \bar{z}) \tag{3.43}$$

and $e^{t\Delta}$ is the inverse of the semigroup generated by the heat equation (namely it is the backwards evolution of P for a time t through the heat semigroup). So, formally one has

$$F(z, \bar{z}, k) = P_b(z, \bar{z}) + O(1/k).$$

3.4.3 Reduction

Consider now the symplectic manifold M obtained by the Marsden–Weinstein reduction procedure from $\mathbb{R}^{2n}$ endowed by the group action ϕ_t, (where ϕ_t denotes again the flow of the linearized Hamiltonian system H_0; in general M is a symplectic orbifold). Namely, assume 1 is a regular value H_0; then

$$M := \frac{H_0^{-1}(1)}{\phi_t}.$$

Moreover, the Hamiltonian (3.38) restricts to M, and its restriction is a Hamiltonian system field with Hamiltonian function $N^t\big|_M$. So the problem of studying the classical dynamics of (3.38) is reduced to the study of the dynamics of $N^t\big|_M$. For the quantum case fix $\hbar = k^{-1}$, then we have the following theorem.

Theorem 4. [Guillemin–Sternberg] *M is naturally endowed with a complex integrable structure and a prequantization bundle L. Furthermore, for any k there exists a canonical isomorphism GS_k from $\mathcal{E}_k$ (see (3.41)) onto the space $\mathcal{H}_k$ of holomorphic sections of $L^{\otimes k}$.*

Remark 8. Here k plays a double role: first it is the inverse of the Planck constant $\hbar$ and secondly it is the number labelling the unperturbed eigenvalue we are studying. Due to this, the limit $k \to \infty$ is actually the semiclassical limit in which $k\hbar$ is fixed.

So the conclusion is that geometric quantization behaves well under symplectic reduction. We are now interested in the behavior of operators when one applies symplectic reduction. We present only the case of interest for the application to resonant systems.

Theorem 5. [Charles [Cha02]] *Consider a Toeplitz operator on the space $\mathcal{B}_k$ with C^∞, compactly supported symbol of the form*

$$f(z, \bar{z}, k) \sim \sum_l k^{-l} f_l(z, \bar{z}) \, ,$$

where the f_l's have compact support, commute with H_0, and the series is asymptotic in the topology of C^∞ functions. Let $(T_f)_k$ be the restriction of T_f to $\mathcal{E}_k$ and f_0^r be the restriction of f_0 to M. Then the operator $\mathrm{GS}_k(T_f)_k\mathrm{GS}_k^{-1}$ is a Toeplitz operator with principal symbol f_0^r.

Corollary 2. *The restriction of $Op^w(N^t)$ to $\mathcal{E}_j$ is a Toeplitz operator with principal symbol*

$$N^t\big|_M \, .$$

3.4.4 The resonant normal form

Remarking that the spectrum of a Toeplitz operator is contained between the maximum and the minimum of its symbol, one immediately has that the spectrum of $Op^w(H)$ has a band structure. Precisely one has

Proposition 5. [BCT03] *Let $\gamma_1 > \frac{1}{2}$. There exists $\hbar_* > 0$ such that for $\hbar \leq \hbar_*$,*

$$Spec(Op^w(H)) \cap (-\infty, \hbar^{\gamma_1}) \subset \bigcup_{E \in Spec(Op^w(H_0))} B(E, \tfrac{1}{3}\omega\hbar)$$

where $B(E, \frac{1}{3}\omega\hbar)$ is the interval $[E - \frac{1}{3}\omega\hbar, E + \frac{1}{3}\omega\hbar]$. Furthermore if $\hbar \leq \hbar_*$, for every eigenvalue E of $Op^w(H_0)$ smaller than $\hbar^\gamma$,

$$\#Spec(Op^w(H)) \cap B(E, \tfrac{1}{3}\omega\hbar) = m(E, \hbar)$$

where $m(E, \hbar)$ is the multiplicity of E.

Then, by the above theory one has that the spectrum in each band can be (approximatively) computed using the normal form. Denote by N_M the restriction of N^t to M, which coincides with the restriction of the classical normal form of order 4 to M, then one has *our main result for the resonant case:*

Theorem 6. [BCT03] *In the considered range of parameters, one has*

$$Spec(Op^w(H)) \cap B(E, \tfrac{1}{3}\omega\hbar) = E + \epsilon^2 Spec(T_{N_M}) \quad \mathrm{mod}\ E^2\left(\frac{\hbar}{E} + E^{1/2}\right).$$
$$(3.44)$$

The first information one can deduce from the above theorem pertains to the extrema of the band and the distribution of the eigenvalues in the band. By standard properties of Toeplitz operators one has

Theorem 7. *Let γ_1, γ_2 be such that $\frac{1}{2} < \gamma_1 < \gamma_2 < 1$. Choose $\hbar_*$ as in proposition (5). For $\hbar \leq \hbar_*$ and for every eigenvalue $E \leq \hbar^{\gamma_1}$ of $\hat{H}$, denote by*

$$E + \lambda_1(E, \hbar) \leq \cdots \leq E + \lambda_{m(E,\hbar)}(E, \hbar)$$

the eigenvalues of $\hat{Q}$ in $B(E, \frac{1}{3}\omega\hbar)$ counted with multiplicity. Then when $\hbar^{\gamma_2} \leq E \leq \hbar^{\gamma_1}$,

$$\lambda_1(E, \hbar) = InfN_M + E^2(O(\hbar/E) + O(E^{\frac{1}{2}}))$$

and similarly

$$\lambda_{m(E,\hbar)}(E, \hbar) = SupN_M + F^2(O(\hbar/E) + O(E^{\frac{1}{2}})) \,.$$

Let f be a C^∞ function on $\mathbb{R}$. Then, when $\hbar^{\gamma_2} \leq E \leq \hbar^{\gamma_1}$,

$$\sum_{i=1}^{m(E,\hbar)} f\left(\lambda_i(E, \hbar)/E^{\frac{3}{2}}\right) = \left(\frac{1}{2\pi\hbar}\right)^{n-1} \int_{\{H=E\}} f\left(N_3(y)/E^{\frac{3}{2}}\right) \mu_E(y)$$
$$+ \left(\frac{E}{\hbar}\right)^{n-1} (O(\hbar/E) + O(E^{\frac{1}{2}}))$$

where μ_E is the Liouville measure of $\{H = E\}$.

References

[Bam95]   D. BAMBUSI, Uniform Nekhoroshev estimates on quantum normal forms, *Nonlinearity* **8** (1995), 93–105.

[BCT03]   D. BAMBUSI, L. CHARLES AND S. TAGLIAFERRO, Resonant normal forms and splitting of degenerate eigenvalues, preprint, 2003.

[BGP99]   D. BAMBUSI, S. GRAFFI, T. PAUL, Normal forms and quantization formulae, *Commun. Math. Phys.* **207** (1999), 173–195.

[BV90]   J. BELLISSARD, M. VITTOT, Heisenberg picture and noncommutative geometry of classical limit in quantum mechanics, *Ann. Inst. H. Poincaré* **52** (1990), 175–235.

[BS91]   F.A. BEREZIN and M.S. SHUBIN, *The Schrödinger Equation*, Kluwer, 1991.

[Cha02]   L. CHARLES, Toeplitz operators and symplectic reduction, preprint, 2002.

[GP87]   S. GRAFFI, T. PAUL, The Schrödinger equation and canonical perturbation theory, *Commun. Math. Phys.* **108** (1987), 25–40.

[Pop00a]   G. POPOV, Invariant tori effective stability and quasimodes with exponentially small error terms I, *Ann. Henri Poincaré* **1** (2000), 223–248.

[Pop00b]   G. POPOV, Invariant tori effective stability and quasimodes with exponentially small error terms II, *Ann. Henri Poincaré* **1** (2000), 249–279.

[Ro87]   D. ROBERT, *Autour de l'approximation semiclassique*, Birkhäuser, Basel, 1987.

[Sjo92]   J. SJÖSTRAND, Semi-excited levels in non-degenerate potential wells, *Asymptotic Analysis* **6** (1992), 29–43.

[Vun98]   S. Vũ Ngọc, Sur le spectre des systemes complètement intégrables semi-classiques avec singularités, thesis, Institut Fourier, 1998.

[VSZB01] Ch. Van Hecke, D. Sadovskii, B. I. Zhilinskiì, V. Boudon, Rotational-vibrational relative equilibria and the structure of quantum energy spectrum of the tetrahedral molecule P_4, *Europ. Phys. J.* **D 17** (2001), 13–35.

[Zhi89]   B. Zhilinskiì, *Chem. Phys.* **137** (1989), 1–13.

4

On the Exit Statistics Theorem of Many-particle Quantum Scattering

D. Dürr and S. Teufel

Mathematisches Institut
Ludwig-Maximilians-Universität München
D-80333 München, Germany
and
Zentrum Mathematik
Technische Universität München
D-80333 München, Germany

Summary. We review the foundations of the scattering formalism for one-particle potential scattering and discuss the generalisation to the simplest case of many non-interacting particles. We point out that the "straight path motion" of the particles, which is achieved in the scattering regime, is at the heart of the crossing statistics of surfaces, which should be thought of as detector surfaces. We prove the relevant version of the many-particle flux across surfaces theorem and discuss what needs to be proven for the foundations of scattering theory in this context.

4.1 Introduction

Quantum mechanical scattering theory is usually about the S-matrix. The operator S maps the so-called in-states α to out-states β. That may seem sufficiently self explanatory as a basic principle since

> *An experimentalist generally prepares a state ... at $t \to -\infty$ and then measures what this state looks like at $t \to +\infty$.* — S. Weinberg in "The quantum theory of fields" [18], Chapter 3.2: The S-Matrix

and

> *The S-matrix $S_{\alpha,\beta}$ is the probability amplitude for the transition $\alpha \to \beta$ ....* — [18] Chapter 3.4: Rates and Cross Sections

so everything seems settled. However the quote continues

> *... but what does this have to do with the transition rates and cross sections measured by experimentalists? ...*
> *... we will give a quick and easy derivation of the main results, actually more a mnemonic than a derivation, with the excuse that (as*

far as I know) no interesting open problems in physics hinge on getting the fine points right regarding these matters. ... — Chapter 3.4: Rates and Cross Sections

The mnemonic recalls that the plane waves in the S-matrix formalism are limits of wave packets, but it does not come to grips with the time-dependent justification of the scattering formalism, in fact it does not connect to the empirical cross section.

We remark aside, that apart from not making contact with the empirical cross section, there is another—though quite related—problem with the mnemonic, which—as is felt by many—can only be settled by interesting new physics: When a particle is scattered by a potential its wave will be spread all over. What accounts then for the fact that a point-particle event is registered at one and only of the detectors? Where did the particle come from which is suddenly manifest in that detector event? This is some facet of the measurement problem of orthodox quantum theory [3, 4]. We shall not say more on that in this paper and refer to [11]. We emphasize however that we shall use Bohmian mechanics for a theoretical description of the cross section—a theory free from the conceptual problems of quantum mechanics.

We immediately jump now to the technical heart of foundations of scattering theory by observing that

$$t \to \pm\infty$$

means the **mathematical** limit of the formulas capturing the **physical** situation (see (4.8) below). Experimentalists prepare and measure states at **large** but **finite** times. They count the number of particles entering the detectors. The physical meaning of the S-matrix derives from being the limit expression of the theoretical formula for the number count. It is moreover immediately clear—once this point of the finiteness of the physical situation has been recognized—that the times at which particles cross the detector surfaces are random. The detector clicks when the particle arrives. That time is random and not fixed by the experimenters. Thus the foundations of quantum mechanical scattering theory become slippery: No observables exist, neither for time measurements nor for position measurements at random times. The question is thus: What are the formulas which theoretically describe the empirical cross section and which result in the appropriate limit in the S-matrix formalism?

In this paper we shall shortly review the simple one-particle potential scattering situation. Apart from discussing the quantum flux we shall introduce Bohmian mechanics, which allows us to capture the theoretical foundations of scattering theory in the most straightforward way. We shall then extend our considerations to multi-particle potential scattering and show why the multi-time flux (which we shall introduce) determines the statistics in this case in terms of a generalized flux across surfaces theorem. The first paper on the flux across surfaces theorem [9] discusses also the multi-particle flux but restricts the computation of statistics to the marginal statistics of one particle only, ignoring thus the most important correlations due to entangled wave functions. Our multi-time analysis deals specifically with entangled wave functions.

4.2 The theoretical cross section

We adopt conventional units in which $\frac{\hbar}{m} = 1$ and recall that the theoretical prediction $\sigma_{\mathbf{k}_0}(\Sigma)$ for the cross section as given by S-matrix theory is

$$\sigma_{k_0}(\Sigma) = 16\pi^4 \int_\Sigma d\omega \, |T(|k_0|\omega, k_0)|^2 \,. \tag{4.1}$$

Here $T = S - I$, where the identity I subtracts the unscattered particles from the scattered beam. As to be explained below, (4.1) is based on a model for a beam of particles. Using heuristic stationary methods, Max Born [7] computed T in the first paper on quantum mechanical scattering theory. We shall recall his argument shortly, since it serves on its own as defining a theoretical cross section.

Consider solutions ψ of the stationary Schrödinger equation with the asymptotics

$$\psi(x) \approx e^{ik_0 \cdot x} + f^{k_0}(\omega)\frac{e^{i|k_0||x|}}{|x|} \quad \text{for } |x| \text{ large} \tag{4.2}$$

and $x = \omega|x|$. In naive scattering theory the first term is regarded as representing an incoming plane wave and the second term as the outgoing scattered wave with angle-dependent amplitude.

Such wave functions can be obtained as solutions of the Lippmann–Schwinger equation

$$\psi(x, k) = e^{ik \cdot x} - \frac{1}{2\pi} \int dy \, \frac{e^{i|k||x-y|}}{|x - y|} \, V(y) \, \psi(y, k) \,. \tag{4.3}$$

The solutions form a complete set, in the sense that an expansion in terms of these generalized eigenfunctions, a so-called generalized Fourier transformation, diagonalizes the continuous spectral part of H. Hence the T-matrix can be expressed in terms of generalized eigenfunctions and one finds (cf. [16]) that

$$T(k, k') = (2\pi)^{-3} \int dx \, e^{-ik \cdot x} \, V(x) \, \psi(x, k') \,. \tag{4.4}$$

Thus the iterative solution of (4.3) yields a perturbative expansion for T, called the Born series.

Moreover, comparing (4.2) and (4.3), expanding the right-hand side of (4.3) in powers of $|x|^{-1}$, we see from the leading term that

$$f^{k_0}(\omega) = -(2\pi)^{-1} \int \mathrm{d}y\, e^{-i|k_0|\omega \cdot y}\, V(y)\, \psi(y, k_0)\,.$$

Thus $f^{k_0}(\omega) = -4\pi^2 T(\omega|k_0|, k_0)$.

We remark that in the so-called naive scattering theory, $f^{k_0}(\omega)$ is called the scattering amplitude since Born's ansatz offers also a heuristic way of defining a cross section. One simply uses the stationary solutions of Schrödinger's equation with the asymptotic behavior (4.2) to obtain the cross section from the quantum probability flux through Σ generated by the scattered wave: The incoming flux has unit density and velocity $v = k_0$. In the outgoing flux generated by $f^{k_0}(\omega)\frac{e^{i|k_0||x|}}{|x|}$ the number of particles crossing an area of size $x^2 \mathrm{d}\omega$ about an angle ω per unit of time is

$$|k_0|(|f^{k_0}(\omega)|^2/|x|^2)|x|^2 \mathrm{d}\omega\,.$$

Normalizing this with respect to the incoming flux suggests the identification of the cross section with

$$\sigma_{k_0}^{\mathrm{naive}}(\Sigma) := \int_{\Sigma} \mathrm{d}\omega\, |f^{k_0}(\omega)|^2 \tag{4.5}$$

in agreement with the above. However, such a heuristic derivation of the formula (4.5) for the cross section, based solely on the stationary picture of a one-particle plane wave function, is unconvincing [8].

4.3 The empirical cross section

Consider a scattering experiment of the most naive kind where one particle is scattered by a potential. In Figure 4.1 we depict a model for such a scattering experiment, where a beam of identical independent particles (defining the ensemble) is shot on a target potential.

The scattering cross section for a potential scattering experiment is measured by the detection rate of particles per solid angle Σ divided by the flux $|j|$ of the incoming beam. ΔT is the total time of duration of the measurement. With $N(\Delta T, R\Sigma)$ denoting the *random* number of particles crossing the surface of the detector located within the solid angle Σ, the empirical distribution is

$$\rho(\Delta T, \Sigma) := \frac{N(\Delta T, R\Sigma)}{\Delta T\, |j|}\,. \tag{4.6}$$

The empirical distribution is a *random variable* on the space of "initial conditions": initial position of the wave packet within the beam, time of creation of wave packet, and also of the *quantum randomness*, encoded in the $|\varphi|^2$

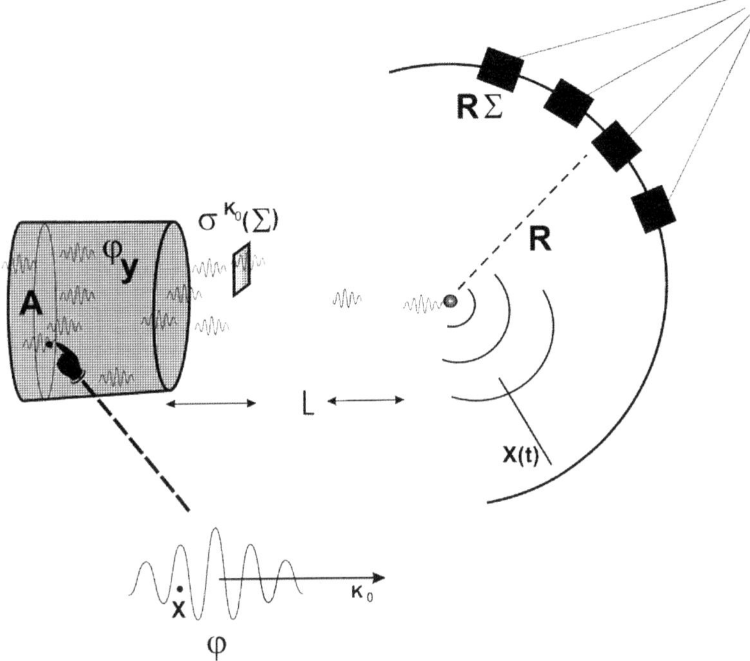

Fig. 4.1. A beam of particles is created in a source far away (distance L) from the scattering center. The particles' waves are all independent from each other. The detectors are a distance R away from the scattering center. In the simplest such models, the wave functions are randomly distributed over the area A of the beam. The particles arrive independently at random times at random positions at the detector surfaces. $\sigma(\Sigma)$ is the cross section, an area which when put in the incident beam is passed by an equal number of particles which per unit of time cross the detector surface defined by the solid angle Σ. The random Bohmian position of the particle within the support of the wave is also depicted as well as its straight Bohmian path $X(t)$ far away from the scattering center.

randomness. It also depends (in fact very much so) on the parameters capturing the physical situation, like the distances L, R and the area A of the beam. The difficult part of this random variable is the dependence on the *quantum randomness*, which, as we shall show, becomes simple in the limit of large distances. We wish to stress that the classical randomness (position of the wave function within the beam, time of creation of the wave function) which arises from the preparation of the beam and which in classical scattering theory is all the randomness there is, adds by virtue of the typical dimensions of the experiment very little to the scattering probabilities in quantum scattering theory (see [11] for more on that).

The goal of scattering theory is to predict the theoretical value of (4.6). The value predicted is (4.1) or if one so wishes (4.5).

What needs to be shown is thus that, in the sense of the law of large numbers,

$$\text{``}\lim_{t\to\pm\infty}\text{''}\ \lim_{\Delta T\to\infty}\rho(\Delta T,\Sigma)=\sigma_{k_0}(\Sigma)\,,\tag{4.7}$$

where the law of large numbers (contained in $\lim_{\Delta T\to\infty}$) will have to be formulated with the measure on the space of the initial conditions. The "$\lim_{t\to\pm\infty}$" refers to large distance limits and limits which make the expression beam-model independent:

$$\text{``}\lim_{t\to\pm\infty}\text{''}=\lim_{|\widehat{\varphi}(k)|^2\to\delta(k-k_0)}\ \lim_{L\to\infty}\ \lim_{|A|\to\infty}\ \lim_{R\to\infty}\,.\tag{4.8}$$

In particular the limit $\lim_{R\to\infty}$ is taken to obtain the "local plane wave" structure (see (4.13)) of the scattered wave, which allows for a particular simple expression for the crossing probability of a particle through the detector surface. For more explanations of the limits see [11, 10].

4.4 The heuristics of quantum randomness

The random number $N(\Delta T,R\Sigma)$ defining (4.6) is the random sum of "independent" single-particle contributions, i.e. it depends on the "trivial" randomness arising from the beam, which is simply ensuring the independence of the single detections in the ensemble for the law of large numbers to hold. Most importantly, however, it depends on the quantum randomness inherent in a single event. We shall from now on focus on the scattering of one single particle and forget the beam. One particle is send towards the scattering center. The question we must then answer is: Which detector clicks? We must answer this question for the real situation where the detectors are a finite distance away from the scattering center. The answer might be complicated but it is that answer of which one can then take the mathematical limit of infinite distances to obtain a simpler looking formula.

Once this question is clear one immediately sees that this question is coarse grained, it already ignores that the time at which the particle is registered is random too. The fundamental question is: *Which detector clicks when?* In other words: What is the distribution for the first exit time and exit position of the particle from the region defined by the detector surfaces (see Figure 4.2).

$$\mathbb{P}^{\varphi}\big(X(T_{\mathrm{e}})\in\mathrm{d}\Sigma,T_{\mathrm{e}}\in\mathrm{d}t\big)=\ ?\tag{4.9}$$

Heuristically it is clear that the probability is given by the quantum flux through the surfaces. The quantum flux is

$$j^{\varphi_t}=\mathrm{Im}\,\varphi_t^*\nabla\varphi_t,$$

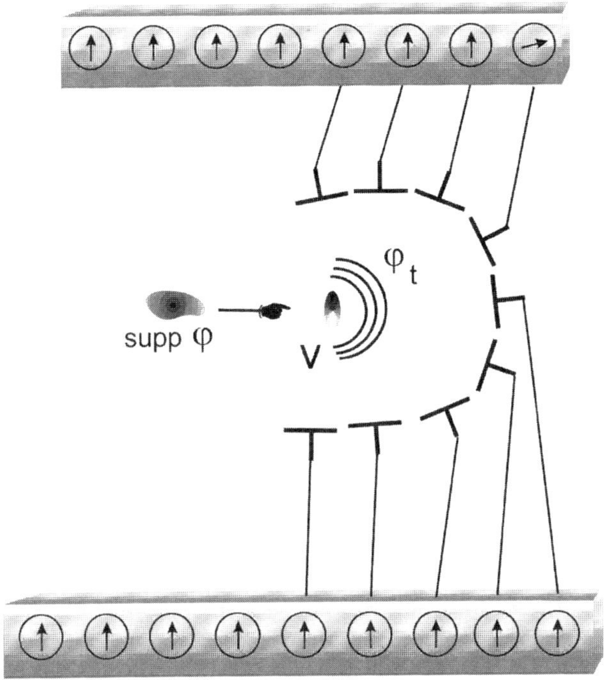

Fig. 4.2. Which detector clicks when? The detection time T_e and position $X_e = X(T_e)$ are *random* exit time and exit position.

and appears in an identity—the so-called quantum flux equation—that holds for any φ_t being a solution of Schrödingers equation:

$$\frac{\partial |\varphi_t|^2}{\partial t} + \operatorname{div} j^{\varphi_t} = 0\,. \tag{4.10}$$

Consider as in Figure 4.3 the escape of a particle initially localized in G through a section dS of the boundary ∂G (we can but need not think of a freely evolving wave). If the surface is far away from the scattering region, it is very suggestive that the probability should be given by the flux integrated against the surface

$$\mathbb{P}^\varphi\big(X(T_e) \in dS, T_e \in dt\big) \approx \lim_{|R|\to\infty} j^{\varphi_t}(R,t) \cdot dS dt\,. \tag{4.11}$$

Based on this heuristic connection the flux across surfaces theorem, which we formulate here in a lax manner, becomes a basic assertion in the foundations of scattering theory [2, 17, 15, 13, 11]. By integrating the flux against the surface integral over all times, we ignore the time at which the particle crosses the surface and we focus merely on the direction in which the particle moves:

Theorem "FAST": Let φ be a (smooth) scattering state, then

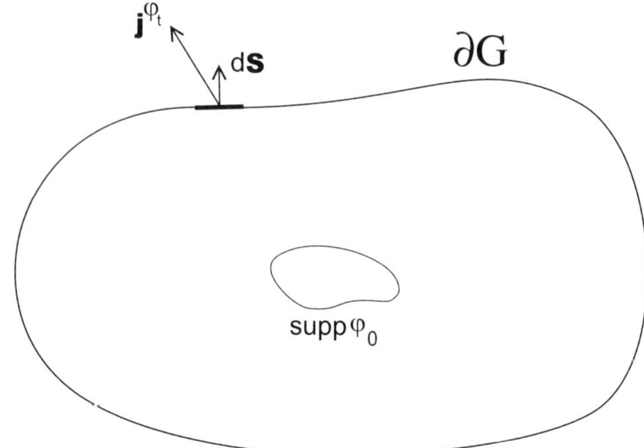

Fig. 4.3. Escape of the particle from the region G. When the boundary ∂G is far from the initial support of the wave function, the exit statistics are approximated by the flux through the surface.

$$\lim_{R \to \infty} \int_0^\infty \mathrm{d}t \int_{R\Sigma} j^{\varphi_t} \cdot \mathrm{d}S = \lim_{R \to \infty} \int_0^\infty \mathrm{d}t \int_{R\Sigma} |j^{\varphi_t} \cdot \mathrm{d}S|$$

(4.12)

$$= \int_{C_\Sigma} \mathrm{d}k \; |\widehat{W_+^* \varphi}(k)|^2 .$$

The heuristics of the FAST is easy to grasp. If we think of a freely evolving wave packet, then its long-time asymptotic (which goes hand in hand with a long-distance asymptotic) is (recall $\frac{\hbar}{m} = 1$)

$$e^{-itH_0} \varphi(x) \approx \frac{e^{\mathrm{i}\frac{x^2}{2t}}}{t^{\frac{3}{2}}} \, \widehat{\varphi}\left(\frac{x}{t}\right) .$$

(4.13)

We call this approximation the local plane wave approximation. It corresponds to a radial outward pointing flux. For scattering states φ of (short range) potential scattering there exists a state φ_{out} moving freely, so that

$$\lim_{t \to \infty} \|e^{-\mathrm{i}Ht}\varphi - e^{-\mathrm{i}H_0 t}\varphi_{\mathrm{out}}\| = 0$$

which leads to the wave operator

$$W_+ := \text{s-}\lim_{t \to \infty} e^{\mathrm{i}Ht} e^{-\mathrm{i}H_0 t}$$

with

$$W_+^* \varphi = \varphi_{\text{out}} .$$

Combining this with (4.13) and computing the flux for this approximation yields that the left-hand side of (4.12) equals the right-hand side of (4.12). We note that the first equality in (4.12) asserts that the flux is outgoing, a condition of vital importance for its interpretation as crossing probability. We shall discuss its importance below. We remark that the further treatment of the right-hand side of (4.12) is more or less standard and becomes upon averaging over the beam statistics essentially (4.1) [1, 11, 10]. That is, given the FAST, the connection with the S-matrix formalism is standard. The cross section is justified in the sense of the law of large numbers, once (4.11) is accepted.

4.5 Bohmian mechanics and the justification of (4.11)

The foregoing discussion is necessarily unprecise since the fundamental objects *exit time* and *exit position* remain undefined: There is no time-dependent position of the particle in quantum theory defining these random variables. In Bohmian mechanics, e.g., [6], when the wave function is φ_t, there is a particle, and the particle moves along a trajectory $X(t)$ determined by the differential equation

$$\frac{\mathrm{d}}{\mathrm{d}t} X(t) = v^{\varphi_t}(X(t)) := \operatorname{Im} \frac{\nabla \varphi_t}{\varphi_t}(X(t)) . \tag{4.14}$$

Its position at time t is randomly distributed according to the probability measure $\mathbb{P}^{\varphi_t}$ having density $\rho_t = |\varphi_t|^2$, see [12].

The continuity equation for the probability transport along the vector field $v^{\varphi_t}(x,t)$ becomes for the particular choice $\rho_t = |\varphi_t|^2$ the quantum flux equation (4.10), which establishes that $|\varphi_t|^2$ is an equivariant density.

Hence the trajectories $X(t, X_0)$ are random trajectories, where the randomness comes from the $\mathbb{P}^\varphi$-distributed random initial position X_0, with φ being the "initial" wave function. Having this, the escape time and position problem (4.9) is readily answered. Define $T_e = \inf\{t \,|\, X(t) \in G^c\}$ and put $X_e = X(T_e)$, then both variables are random variables on the space of initial positions of the particle and $\mathbb{P}^\varphi(\{X \,|\, T_e(X) \in \mathrm{d}t, \, X(T_e(X), X) \in \mathrm{d}S\})$ is clearly the exit distribution we are looking for. Note also, that we may now specify rigorously the probability space on which the empirical distribution (4.6) is naturally defined, and we furthermore have the measure, with which the law of large numbers (4.7) can be proven.

We explain now the connection of this exit probability with the flux. Consider some possible exit scenarios of the particle as in Figure 4.4. We introduce the random variables *number of crossings*

$$N(\mathrm{d}S, \mathrm{d}t) := N_+(\mathrm{d}S, \mathrm{d}t) + N_-(\mathrm{d}S, \mathrm{d}t)$$

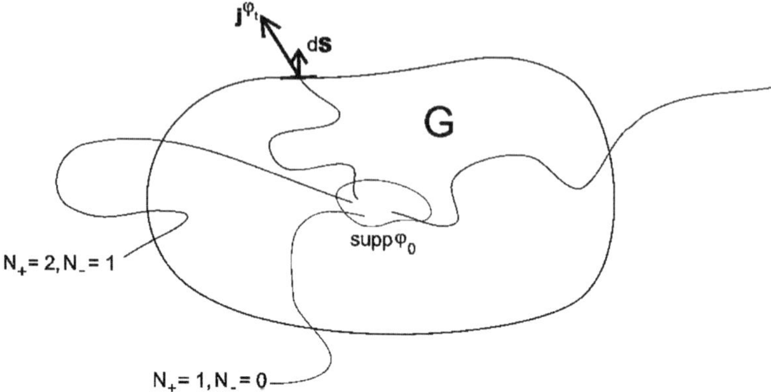

Fig. 4.4. Signed number of crossings of possible trajectories through the boundary of the region G.

and *number of signed crossings*

$$N_s(\mathrm{d}S, \mathrm{d}t) := N_+(\mathrm{d}S, \mathrm{d}t) - N_-(\mathrm{d}S, \mathrm{d}t)\,,$$

where $N_\pm(\mathrm{d}S, \mathrm{d}t)$ are the number of outward, resp. inward, crossings. Their expectations are readily computed in the usual statistical mechanics manner: For a crossing of $\mathrm{d}S$ in the time interval $(t, t + \mathrm{d}t)$ to occur, the particle has to be in a cylinder (Boltzmann collision cylinder) of size $|v^{\varphi_t} \cdot \mathrm{d}S\,\mathrm{d}t|$ at time t. Thus

$$\mathbb{E}^\varphi(N(\mathrm{d}S, \mathrm{d}t)) = |\varphi_t|^2\,|v^{\varphi_t} \cdot \mathrm{d}S|\,\mathrm{d}t = |j^{\varphi_t} \cdot \mathrm{d}S|\,\mathrm{d}t$$

and

$$\mathbb{E}^\varphi(N_s(\mathrm{d}S, \mathrm{d}t)) = j^{\varphi_t} \cdot \mathrm{d}S\,\mathrm{d}t\,. \tag{4.15}$$

Under the condition that the flux is positive for all times through the boundary of G (a condition which needs to be proven, and which is asserted in the first equality of (4.12)) every trajectory crosses the boundary of G at most once. Hence

$$\mathbb{E}^\varphi(N(\mathrm{d}S, \mathrm{d}t)) = \mathbb{E}^\varphi(N_s(\mathrm{d}S, \mathrm{d}t))$$
$$= 0 \cdot \mathbb{P}^\varphi(T_e \notin \mathrm{d}t \text{ or } X_e \notin \mathrm{d}S) + 1 \cdot \mathbb{P}^\varphi(X_e \in \mathrm{d}S \text{ and } T_e \in \mathrm{d}t)\,.$$

In that particular situation the exit probability is thus

$$\mathbb{P}^\varphi(X_e \in \mathrm{d}S \text{ and } T_e \in \mathrm{d}t) = j^{\varphi_t} \cdot \mathrm{d}S\,\mathrm{d}t\,. \tag{4.16}$$

This and (4.12) are at the basis of quantum mechanical scattering theory for single-particle potential scattering.

4.6 Multi-time distributions for many particles

We extend the foregoing to the case of many-particle scattering. We shall discuss some of the main steps, which need to be filled with rigorous mathematics in future works. For simplicity we consider the free case where the particles are guided by an entangled wave function, but they do not interact via a potential term in the Hamiltonian with each other. However, the following naturally generalizes to interacting particles by replacing the wave function φ by its free outgoing asymptote $\varphi_{\text{out}} = W_+^* \, \varphi$. While Bohmian mechanics naturally extends to many particles (see (4.19) below), one sees immediately that our task of getting our hands on the exit statistics for many particles is nevertheless nontrivial, since every particle has its own exit time and position. I.e. we need to handle

$$\mathbb{P}^\varphi\left(T_{\mathrm{e}}^{(1)} \in \mathrm{d}t^{(1)}, X^{(1)}(T_{\mathrm{e}}^{(1)}) \in \mathrm{d}S^{(1)}, \dots, T_{\mathrm{e}}^{(n)} \in \mathrm{d}t^{(n)}, X^{(n)}(T_{\mathrm{e}}^{(n)}) \in \mathrm{d}S^{(n)}\right).$$
$$(4.17)$$

To apply the statistical mechanics argument which we used in the last section to compute the crossing probability, the multi-time position distribution is needed

$$\mathbb{P}^\varphi\left(X^{(1)}(t^{(1)}) \in \mathrm{d}x^{(1)}, \dots, X^{(n)}(t^{(n)}) \in \mathrm{d}x^{(n)}\right) \tag{4.18}$$
$$= \rho(x^{(1)}, t^{(1)}, \dots, x^{(n)}, t^{(n)}) \, \mathrm{d}x^{(1)} \dots \mathrm{d}x^{(n)},$$

which in general will not be a simple functional of the wave function. We will show that in the scattering regime, when the wave approaches the local plane wave structure, this multi-time position distribution can be computed and the exit statistics are in fact given by a particular multi-time flux form. To our best knowledge, this observation is new. The single-time multi-particle flux has been used in [9] to compute exist statistics, necessarily ignoring particle correlations.

For ease of notation we consider two particles with positions X, Y and wave function $\varphi(x, y, t)$. The Bohmian law of motion is

$$\dot{X}(t) = v_t^x(X(t), Y(t)) = \mathrm{Im} \left. \frac{\nabla_x \varphi(x, y, t)}{\varphi(x, y, t)} \right|_{x=X(t), \, y=Y(t)}, \tag{4.19}$$

$$\dot{Y}(t) = v_t^y(X(t), Y(t)) = \mathrm{Im} \left. \frac{\nabla_y \varphi(x, y, t)}{\varphi(x, y, t)} \right|_{x=X(t), \, y=Y(t)}, \tag{4.20}$$

$$\mathrm{i}\partial_t \varphi(x, y, t) = -\tfrac{1}{2}\left(\Delta_x + \Delta_y\right)\varphi(x, y, t). \tag{4.21}$$

With $H = H_x + H_y = -\tfrac{1}{2}\left(\Delta_x + \Delta_y\right)$ we can easily produce a two-times wave function by the appropriate action of the single-particle Hamiltonians through

$$\varphi(x,t,y,s) := \left(e^{-iH_x t}e^{-iH_y s}\varphi_0\right)(x,y)\,, \tag{4.22}$$

which reduces to the usual single-time wave function for $t = s$, because the Hamiltonians H_x and H_y commute. Hence one could as well include single-particle potentials into H_x and H_y. While the definition of $\varphi(x,t,y,s)$ seems very natural at first sight, note that the physical meaning of $|\varphi(x,t,y,s)|^2$ is not at all obvious. To get our hands on this question, let

$$\Phi_t(x,y) = \left(\Phi_t^x(x,y), \Phi_t^y(x,y)\right) = \left(X(t,x,y), Y(t,x,y)\right)$$

be the Bohmian flow along the vector field given by (4.19) transporting the initial values x, y along the Bohmian trajectories to values at time t and let

$$\Phi_{t,s}(x,y) = \left(\Phi_t^x(x,y), \Phi_s^y(x,y)\right) = \left(X(t,x,y), Y(s,x,y)\right)$$

be the two-times Bohmian flow. Observe that

$$\partial_t \Phi_{t,s}(x,y) = \left(\partial_t \Phi_t^x(x,y), 0\right) = \left(v_t^x\left(\Phi_t(x,y)\right), 0\right), \tag{4.23}$$

$$\partial_s \Phi_{t,s}(x,y) = \left(0, \partial_s \Phi_s^y(x,y)\right) = \left(0, v_s^y\left(\Phi_s(x,y)\right)\right). \tag{4.24}$$

From the definition of the multi-time wave function (4.22) it follows in the same way as in the single-time case that

$$\partial_t|\varphi(x,t,y,s)|^2 = -\nabla_x \cdot \mathrm{Im}\left(\varphi(x,t,y,s)^* \nabla_x \varphi(x,t,y,s)\right),$$
$$\partial_s|\varphi(x,t,y,s)|^2 = -\nabla_y \cdot \mathrm{Im}\left(\varphi(x,t,y,s)^* \nabla_y \varphi(x,t,y,s)\right), \tag{4.25}$$

which leads us to define a multi-time velocity field:

$$v_{t,s}^x(x,y) = \mathrm{Im}\,\frac{\nabla_x \varphi(x,t,y,s)}{\varphi(x,t,y,s)} \tag{4.26}$$

if $\varphi(x,t,y,s) \neq 0$ and $v_{t,s}^x(x,y) = 0$ if $\varphi(x,t,y,s) = 0$ and analogously for $v_{t,s}^y(x,y)$.

We show now, that under certain conditions there exists a two-times continuity equation for a two-times density $\rho(x,t,y,s)$. We start with the definition, setting $\rho(x,0,y,0) = \rho(x,y)$,

$$\mathbb{E}^\varphi\left(f\left(X(t), Y(s)\right)\right) = \int \mathrm{d}x\mathrm{d}y\, f\left(\Phi_{t,s}(x,y)\right)\rho(x,y)$$

$$=: \int \mathrm{d}x\mathrm{d}y\, f(x,y)\rho(x,t,y,s)\,, \tag{4.27}$$

where f varies in a suitable class of test functions. Next differentiate the equation with respect to t, respectively s. This yields in the second equality

$$\partial_t \int \mathrm{d}x\mathrm{d}y\, f\left(\Phi_{t,s}(x,y)\right)\rho(x,y)$$

$$= \int \mathrm{d}x\mathrm{d}y\, \nabla_{(1)}f\left(\Phi_{t,s}(x,y)\right) \cdot v_t^x\left(\Phi_t(x,y)\right)\rho(x,y)$$

$$= \int \mathrm{d}x\mathrm{d}y\, f(x,y)\partial_t\rho(x,t,y,s)\,, \tag{4.28}$$

and similarly for differentiation with respect to s. Here $\nabla_{(1)}$ denotes the gradient with respect to the first argument. If the following "multi-time independence" condition

$$v_t^x\big(\varPhi_t(x,y)\big) = v_{t,s}^x\big(\varPhi_t^x(x,y),\varPhi_s^y(x,y)\big),$$
$$v_t^y\big(\varPhi_t(x,y)\big) = v_{t,s}^y\big(\varPhi_t^x(x,y),\varPhi_s^y(x,y)\big) \tag{4.29}$$

is satisfied, we can replace $v_t^x\big(\varPhi_t(x,y)\big), v_t^y\big(\varPhi_t(x,y)\big)$ in (4.28) by

$$v_{t,s}^x\big(\varPhi_t^x(x,y),\varPhi_s^y(x,y)\big)\,.$$

Using definition (4.27) followed by partial integration yields for the second integral in (4.28)

$$\int \mathrm{d}x\mathrm{d}y\,\nabla_{(1)}f\big(\varPhi_{t,s}(x,y)\big)\cdot v_t^x\big(\varPhi_t(x,y)\big)\rho(x,y)$$
$$= \int \mathrm{d}x\mathrm{d}y\,\nabla_{(1)}f\big(\varPhi_{t,s}(x,y)\big)\cdot v_{t,s}^x\big(\varPhi_{t,s}(x,y)\big)\rho(x,y)$$
$$\overset{(4.28)}{=} \int \mathrm{d}x\mathrm{d}y\,\nabla_{(1)}f\big(x,y\big)\cdot v_{t,s}^x\big(x,y\big)\rho(x,t,y,s)$$
$$= -\int \mathrm{d}x\mathrm{d}y\,f(x,y)\,\nabla_x\cdot\big(v_{t,s}^x(x,y)\rho(x,t,y,s)\big)\,.$$

From this and (4.28) we may conclude, repeating the same for the s-differentiation, the two-times continuity equation

$$\partial_t\rho(x,t,y,s) = -\nabla_x\cdot\big(v_{t,s}^x(x,y)\rho(x,t,y,s)\big),$$
$$\partial_s\rho(x,t,y,s) = -\nabla_y\cdot\big(v_{t,s}^y(x,y)\rho(x,t,y,s)\big)\,. \tag{4.30}$$

Comparing this with (4.25) we see that $\rho(x,t,y,s) = |\varphi(x,t,y,s)|^2$ is equivariant. All this depends crucially on the "multi-time independence" condition (4.29). It is easy to see that the condition is satisfied if the wave function is a product wave function. But that is uninteresting. The condition can be expected to be also approximately satisfied when the wave function attains the local plane wave structure

$$\varphi(x,t,y,s) \approx \frac{e^{\mathrm{i}\frac{x^2}{2t}}}{t^{\frac{3}{2}}}\frac{e^{\mathrm{i}\frac{y^2}{2s}}}{s^{\frac{3}{2}}}\,\widehat{\varphi}\Big(\frac{x}{t},\frac{y}{s}\Big) \tag{4.31}$$

of an outgoing scattering state at large times (see next section). In this case the trajectories are approximately straight lines and the velocity of particle X does not change if particle Y is moved along its straight path and vice versa. We remark that the local plane wave structure is preserved under multi-time evolution (as it is preserved under single-time evolution). Thus in the scattering regime condition (4.29) holds true and we conclude that in this regime the two-times wave function (4.22) yields the two-times joint distribution $\rho(x, t, y, s) = |\varphi(x, t, y, s)|^2$ for the positions of the two particles. Hence, approximately, we have that

$$\mathbb{P}^\varphi\left(X(t) \in \Lambda_1 \text{ and } Y(s) \in \Lambda_2\right) \approx \int_{\Lambda_1} dx \int_{\Lambda_2} dy \, |\varphi(x, t, y, s)|^2 \,.$$

Moreover we have in that regime single crossings only. We can thus compute the exit statistics in the scattering regime as before (the Boltzmann collision cylinder argument) but now using the two-times density $|\varphi(x, t, y, s)|^2$ and the approximate straight path velocities

$$v_{t,s}^x(x, y) \approx \frac{x}{t} \ , \ v_{t,s}^y(x, y) \approx \frac{y}{s} \tag{4.32}$$

This way one obtains

$$\mathbb{P}^\varphi(T_e^x \in dt, T_e^y \in ds, X(T_e^x) \in dS^x, Y(T_e^y) \in dS^y) \tag{4.33}$$
$$\approx |\hat{\varphi}(\frac{x}{t}, \frac{y}{s})|^2 \left(\frac{x}{t} \cdot dS^x\right) \left(\frac{y}{s} \cdot dS^y\right) dt \, ds$$
$$\approx: j^{\mathrm{SP}}(x, t, y, s) \cdot (dS^x \otimes dS^y) \, dt \, ds \,,$$

where the two-times "straight paths" flux form $j^{\mathrm{SP}}(x, t, y, s)$ is the straight path approximation to the multi-time flux form

$$j(x, t, y, s) := |\varphi(x, t, y, s)|^2 \, v_{t,s}^x(x, y) \otimes v_{t,s}^y(x, y) \,. \tag{4.34}$$

It is remarkable and relevant for its meaning in the foundations of scattering theory that this *unmeasured* Bohmian joint probability is in this particular situation the same as the measured probability, which is in general not true for joint probabilities [5]. Measurements lead—in the language of orthodox quantum theory—to a collapse of the wave function, which in the local plane wave approximation however does not have any effect on the trajectory of the other particles. In the two-particles case the collapse (due to the detection of one particle) picks out simply the rightly correlated pair, which in fact can be EPR correlated pairs.

The N-particle multi-time flux (4.34) as well as the N-particle single-time flux have taken alone no significance for the description of scattering (in contrast to the one-particle situation), while the crossing probabilities (4.33) of course do. We shall in the next section compute the value of the right-hand side of (4.33), which is the usual scattering into cones (in momentum space) formula.

4.7 The exit statistics theorem for N particles

We abbreviate the joint exit time-exit position distribution for N particles through a sphere of radius R as

$$\mathbb{P}^{\varphi}(\mathrm{d}t_1 \ldots \mathrm{d}t_N \, \mathrm{d}S_1 \ldots \mathrm{d}S_N)$$
$$:= \mathbb{P}^{\varphi}(X_1(T_{1\mathrm{e}}) \in \mathrm{d}S_1, T_{1\mathrm{e}} \in \mathrm{d}t_1, \ldots, X_N(T_{N\mathrm{e}}) \in \mathrm{d}S_1, T_{N\mathrm{e}} \in \mathrm{d}t_N),$$

where we recall that $T_{n\mathrm{e}}$ is the first exit time of the nth particle through the sphere and $\mathrm{d}S_n$ an infinitesimal surface element on this sphere. Neglecting the possibility of clustering, the generalization of the flux-across surfaces theorem of potential scattering then becomes the following conjecture.

Exit Statistics Theorem: *Let φ be a (smooth) scattering state of an N-body Hamiltonian H at time $t = 0$; then for any $-\infty < T < \infty$,*

$$\lim_{R \to \infty} \int_T^{\infty} \cdots \int_T^{\infty} \int_{R\Sigma_1} \cdots \int_{R\Sigma_N} \mathbb{P}^{\varphi}(\mathrm{d}t_1 \ldots \mathrm{d}t_N \, \mathrm{d}S_1 \ldots \mathrm{d}S_N)$$

$$= \lim_{R \to \infty} \int_T^{\infty} \mathrm{d}t_1 \cdots \int_T^{\infty} \mathrm{d}t_N \int_{R\Sigma_1} \cdots$$

$$\cdots \int_{R\Sigma_N} j^{\varphi_{\mathrm{out},\mathrm{sp}}}(x_1, \ldots, x_N, t_1, \ldots, t_N)(\mathrm{d}S_1 \otimes \cdots \otimes \mathrm{d}S_N)$$

$$= \int_{C_{\Sigma_1}} \mathrm{d}k_1^3 \cdots \int_{C_{\Sigma_N}} \mathrm{d}k_N^3 \, |\widehat{\varphi}_{\mathrm{out}}(k_1, \ldots, k_N)|^2. \tag{4.35}$$

Recall that $\varphi_{\mathrm{out}} = W_+^* \varphi$ and that

$$\varphi_{\mathrm{out}}(t_1, \ldots, t_N) = \mathrm{e}^{\mathrm{i}\Delta_{x_1} t_1} \cdots \mathrm{e}^{\mathrm{i}\Delta_{x_N} t_N} \varphi_{\mathrm{out}}$$

evolves according to the free multi-time evolution.

The theorem provides a precise connection between the joint distribution of the measured exit positions of N scattered particles (the first expression in (4.35)) and the empirical formula for this quantity in terms of the Fourier transform of the outgoing wave (the last expression in (4.35)). A rigorous proof of this connection seems to involve necessarily a multi-time formulation of the quantum mechanics in the scattering regime in the sense of the intermediate expression in (4.35). Notice that the first equality in (4.35) is, as discussed in the previous section, the highly nontrivial part to prove. More precisely, one needs to establish (4.33) rigorously and with error estimates which are integrable in the sense of (4.35). The second equality in (4.35) is an easy computation, with which we shall conclude the paper. We shall first remind the reader of the local plane wave structure which approximates the scattering state and which is presumably crucial for the proof of the theorem.

Since $|\widehat{\varphi}_{\mathrm{out}}(k)|$ is invariant under the free time-evolution we can choose without loss of generality $T \geq 1$. To shorten notation let us introduce the configuration variables $\overline{x} = (x_1, \ldots, x_N)$ and $\overline{t} = (t_1, \ldots, t_N)$. Then

$$\varphi_{\mathrm{out}}(\overline{x}, \overline{t}) = \left(\mathrm{e}^{\mathrm{i}\Delta_{x_1} t_1} \cdots \mathrm{e}^{\mathrm{i}\Delta_{x_N} t_N}\right) \varphi_{\mathrm{out}}(\overline{x})$$

$$= \int_{\mathbb{R}^3} \mathrm{d}y_1 \cdots \int_{\mathbb{R}^3} \mathrm{d}y_N \, \frac{\mathrm{e}^{\mathrm{i}\frac{|x_1 - y_1|^2}{2t_1}}}{(2\pi \mathrm{i}t_1)^{\frac{3}{2}}} \cdots \frac{\mathrm{e}^{\mathrm{i}\frac{|x_N - y_N|^2}{2t_N}}}{(2\pi \mathrm{i}t_N)^{\frac{3}{2}}} \varphi_{\mathrm{out}}(y_1, \ldots, y_N),$$

where here and in the following φ_{out} without a time-argument means always $\varphi_{\mathrm{out}}(\bar{t}=0)$. Expanding every factor in the integrand as

$$\mathrm{e}^{\mathrm{i}\frac{|x_n-y_n|^2}{2t_n}} = \mathrm{e}^{\mathrm{i}\frac{|x_n|^2}{2t_n}}\,\mathrm{e}^{-\mathrm{i}\frac{x_n\cdot y_n}{t_n}} + \mathrm{e}^{\mathrm{i}\frac{|x_n|^2}{2t_n}}\,\mathrm{e}^{-\mathrm{i}\frac{x_n\cdot y_n}{t_n}}\left(\mathrm{e}^{\mathrm{i}\frac{|y_n|^2}{2t_n}}-1\right),$$

one obtains

$$\varphi_{\mathrm{out}}(\bar{x},\bar{t}) = \frac{\mathrm{e}^{\mathrm{i}\frac{|x_1|^2}{2t_1}}}{(it_1)^{\frac{3}{2}}}\cdots\frac{\mathrm{e}^{\mathrm{i}\frac{|x_N|^2}{2t_N}}}{(it_N)^{\frac{3}{2}}}\,\widehat{\varphi}_{\mathrm{out}}\left(\frac{x_1}{t_1},\ldots,\frac{x_N}{t_N}\right) + R(\bar{x},\bar{t}), \qquad (4.36)$$

where every term in the sum R has at least one factor of the form

$$\left(\mathrm{e}^{\mathrm{i}\frac{|y_n|^2}{2t_n}}-1\right)$$

in the integrand. Under appropriate assumptions on φ_{out} it is now easy to get estimates on the remainder term $R(\bar{x},\bar{t})$ for large t_n by stationary phase methods. For details we refer to [14]. In particular the remainder term does not contribute to the time integrals in (4.35).

Neglecting R we obtain from (4.36) for the nth component of the velocity

$$v_{\bar{t}}^n(\bar{x}) = \frac{x_n}{t_n} + \frac{1}{t_n}\,\mathrm{Im}\,\frac{\nabla_n\widehat{\varphi}_{\mathrm{out}}\left(\frac{x_1}{t_1},\ldots,\frac{x_N}{t_N}\right)}{\widehat{\varphi}_{\mathrm{out}}\left(\frac{x_1}{t_1},\ldots,\frac{x_N}{t_N}\right)}, \qquad (4.37)$$

of which we only need the first term (the straight path velocity) and for the density

$$|\varphi_{\mathrm{out}}(\bar{x},\bar{t})|^2 = \frac{1}{t_1^3\cdots t_N^3}\left|\widehat{\varphi}_{\mathrm{out}}\left(\frac{x_1}{t_1},\ldots,\frac{x_N}{t_N}\right)\right|^2.$$

Using $x_n\cdot\mathrm{d}S_n = |x_n|R^2\mathrm{d}\omega_n = R^3\mathrm{d}\omega_n$, where $\mathrm{d}\omega$ denotes Lebesgue measure on the unit sphere $S^2\subset\mathbb{R}^3$, we now conclude with the computation of the second equality of (4.35):

$$\lim_{R\to\infty}\int_T^\infty\mathrm{d}t_1\cdots\int_T^\infty\mathrm{d}t_N\int_{R\Sigma_1}\cdots\int_{R\Sigma_N} j^{\varphi_{\mathrm{out,sp}}}(\bar{x},\bar{t})\cdot(\mathrm{d}S_1\otimes\ldots\otimes\mathrm{d}S_N)$$

$$= \lim_{R\to\infty}\int_T^\infty\mathrm{d}t_1\cdots\int_T^\infty\mathrm{d}t_N\int_{R\Sigma_1}\cdots\int_{R\Sigma_N}\frac{\left|\widehat{\varphi}_{\mathrm{out}}\left(\frac{R\omega_1}{t_1},\ldots,\frac{R\omega_N}{t_N}\right)\right|^2}{t_1^4\cdots t_N^4}R^{3N}$$

$$\mathrm{d}\omega_1\cdots\mathrm{d}\omega_N$$

$$= \lim_{R\to\infty}\int_0^{\frac{R}{T}}\mathrm{d}|k_1|\cdots\int_0^{\frac{R}{T}}\mathrm{d}|k_N|\int_{R\Sigma_1}\cdots\int_{R\Sigma_N}|\widehat{\varphi}_{\mathrm{out}}(k_1,\ldots,k_N)|^2$$

$$|k_1|^2\cdots|k_N|^2\,\mathrm{d}\omega_1\cdots\mathrm{d}\omega_N$$

$$= \int_{C_{\Sigma_1}}\mathrm{d}k_1^3\cdots\int_{C_{\Sigma_N}}\mathrm{d}k_N^3\,|\widehat{\varphi}_{\mathrm{out}}(k_1,\ldots,k_N)|^2.$$

In the above computation we substituted $k_n = \frac{x_n}{t_n}$, which, in particular, gives $\mathrm{d}t_n = -t_n^2 R^{-1}\mathrm{d}|k_n|$ and $R/t_n = |k_n|$.

4.8 Conclusion

For the first time we formulate the connection between the joint distribution of the measured exit positions of N scattered particles and the empirical formula for this quantity in terms of the Fourier transform of the outgoing wave. While in the case of potential scattering for a single particle the distribution of the measured exit position can be formulated, at least heuristically, in terms of the quantum flux, this is no longer true for the joint distribution of N particles. In the case of N-particle scattering even the definition of the relevant distribution is not possible within orthodox quantum mechanics. Therefore we use the Bohmian trajectories of the particles to define the distribution of exit positions and times. The flux-across-surfaces theorem for N particles then connects this fundamental joint distribution with the empirical formulas of quantum mechanics. While a completely rigorous proof of the flux-across-surfaces theorem for N particles seems a challenging task, we sketched a possible argument and showed that a multi-time formulation of the quantum mechanics in the scattering regime should play a crucial role in this program.

References

1. W.O. Amrein, J.M. Jauch and K.B. Sinha, *Scattering Theory in Quantum Mechanics*, W.A. Benjamin, Inc., Reading, Massachusetts, 1977.
2. W.O. Amrein and D.B. Pearson, Flux and scattering into cones for long range and singular potentials, *Journal of Physics A* **30** (1997), 5361–5379.
3. J.S. Bell, *Speakable and Unspeakable in Quantum Mechanics*, Cambridge University Press, Cambridge, 1987.
4. J.S. Bell, Against measurement, *Physics World* **3** (1990), 33–40.
5. K. Berndl, D. Dürr, S. Goldstein, G. Peruzzi, and N. Zanghì, On the global existence of Bohmian mechanics, *Comm. Math. Phys.* **173** (1995), 647–673.
6. D. Bohm and B.J. Hiley, *The Undivided Universe: An Ontological Interpretation of Quantum Theory*, Routledge, Chapman and Hall, London, 1993.
   D. Dürr, S. Goldstein and N. Zanghì, Bohmian mechanics as the foundations of quantum mechanics, in *Bohmian Mechanics and Quantum Theory: An Appraisal*, edited by J. Cushing, A. Fine and S. Goldstein, Boston Studies in the Philosophy of Science Vol. 184, Kluwer Academic Press, 1996.
7. M. Born. Quantenmechanik der Stoßvorgänge, *Zeitschrift für Physik* **38** (1926), 803–827.
8. C. Cohen-Tannoudji, B. Diu and F. Laloë, *Quantenmechanik. Teil 2*, de Gruyter, Berlin, 1997; R. Haag: *Local Quantum Physics*, 2nd edition, Springer-Verlag, Berlin, 1996.

9. J.-M. Combes, R.G. Newton and R. Shtokhamer, Scattering into cones and flux across surfaces, *Phys. Rev. D* **11** (1975), 366–372.
10. D. Dürr, S. Goldstein, T. Moser and N. Zanghì, in preparation.
11. D. Dürr, S. Goldstein, S. Teufel and N. Zanghì, Scattering theory from microscopic first principles, *Physica A* **279** (2000), 416–431.
12. D. Dürr, S. Goldstein and N. Zanghì, Quantum equilibrium and the origin of absolute uncertainty, *J. Stat. Phys.* **67** (1992), 843–907.
13. G. Dell'Antonio and G. Panati, Zero-energy resonances and the flux-across-surfaces theorem, preprint, ArXive math-ph/0110034 (2001).
14. D. Dürr and S. Teufel, On the role of the flux in scattering theory, in *Stochastic Processes, Physics and Geometry: New Interplays I*, Holden et al., eds. Conference Proceedings of the Canadian Mathematical Society, 2001.
15. G. Panati and A. Teta, The flux-across-surfaces theorem for a point interaction Hamiltonian, in *Stochastic Processes, Physics and Geometry: New Interplays I*, Holden et al., eds. Conference Proceedings of the Canadian Mathematical Society, 2001.
16. M. Reed and B. Simon, *Methods of Modern Mathematical Physics III: Scattering Theory*, Academic Press, New York, 1978.
17. S. Teufel, D. Dürr and K. Münch-Berndl, The flux-across-surfaces theorem for short range potentials and wave functions without energy cutoffs, J. Math. Phys. **40** (1999), 1901–1922.
18. S. Weinberg, *The Quantum Theory of Fields*, Cambridge University Press, 1995.

5

Two-scale Wigner Measures and the Landau–Zener Formulas

C. Fermanian Kammerer and P. Gérard

Université de Cergy-Pontoise
2 av. Adolphe Chauvin
BP 222 Pontoise
F-95302 Cergy Pontoise Cedex, France
and
Laboratoire de Mathématiques d'Orsay
Université Paris-Sud - Bât 425
F–91405 Orsay Cedex, France
CNRS UMR 8628

5.1 Introduction

We consider the system of evolution equations

$$i\varepsilon \frac{\partial \psi^\varepsilon}{\partial t} = \mathrm{op}_\varepsilon (H)\psi^\varepsilon,$$

where ε is a small parameter, $\psi^\varepsilon = \psi^\varepsilon(t,x)$ a vector-valued bounded family in $L^2(\mathbb{R}^d)$, $H = H(t,x,\xi)$ a matrix-valued Hamiltonian. The variable x denotes the position variable and ξ the momentum. We use Weyl quantization:

$$\mathrm{op}_\varepsilon (a) = \int_{\mathbb{R}^d \times \mathbb{R}^d} e^{i(x-y)\cdot\xi} a\left(\frac{x+y}{2}, \varepsilon\xi\right) f(y) \frac{dy \, d\xi}{(2\pi)^d}.$$

We are concerned with the description as ε goes to 0 of the Wigner transform $W\psi^\varepsilon(t,x,\xi)$ of the family (ψ^ε),

$$W\psi^\varepsilon(t,x,\xi) = \int_{\mathbb{R}^d} e^{iv\cdot\xi} \psi^\varepsilon(t, x - \frac{v}{2}) \otimes \psi^\varepsilon(t, x + \varepsilon\frac{v}{2}) \frac{dv}{(2\pi)^d}.$$

More precisely, for all reasonable a, we want to describe the asymptotic behavior of

$$(\mathrm{op}_\varepsilon (a)\psi^\varepsilon \mid \psi^\varepsilon)_{L^2} = \int_{\mathbb{R}^d \times \mathbb{R}^d} a(x,\xi) W\psi^\varepsilon(t,x,\xi) \, dx \, d\xi.$$

It is well known that the limit points of $W\psi^\varepsilon$ are positive matrix-valued measures μ (see [17], [11]), called *semi-classical measure* or *Wigner measure* of the family (ψ^ε).

If H is scalar, $\mu = \mu_t$ is a scalar measure solution of

$$\partial_t \mu + \{H, \mu\} = 0$$

where

$$\{f, g\} := \nabla_\xi f \cdot \nabla_x g - \nabla_x f \cdot \nabla_\xi g.$$

In other words, the measure μ propagates along the classical Hamiltonian trajectories.

If H has N eigenvalues of constant multiplicity $E_1, \cdots, E_N$, then

$$\mu = \underset{j}{\Sigma}\, \mu_j \Pi_j$$

where Π_j is the spectral projector associated to E_j and μ_j satisfies

$$\partial_t \mu_j + \{E_j, \mu_j\} = 0.$$

We want to discuss what happens when H displays eigenvalues crossing. Of typical interest is the situation where H is a two-by-two matrix of the form

$$H = k\,\mathrm{Id} + \begin{pmatrix} p_1 & p_2 + ip_3 \\ p_2 - ip_3 & -p_1 \end{pmatrix}.$$

The eigenvalues of H are $E_\pm = k \pm |p|$. They cross above the codimension 4 subset of the phase space

$$S = \{p_1 = p_2 = p_3 = \tau + k = 0\} \subset T^*(\mathbb{R}_t \times \mathbb{R}_x^d).$$

Under appropriate assumptions that we shall state below, classical trajectories of $E_\pm$ can be continued through S. Then occurs the following problem:

Problem: Assume that the incoming Wigner measure is supported on the classical trajectories of one mode (say E_+) which hit S. How can the splitting of the outgoing Wigner measure on both modes be described?

Such problems appear in different areas of Mathematical Physics.
a) The crossings in Born–Oppenheimer approximation have been studied by Hagedorn ([12], [13]) and Hagedorn and Joye ([14], [15]). In their works, the Hamiltonian H is of the form

$$H(t, x, \xi) = |\xi|^2 + \begin{pmatrix} q(x) + p_1(x) & p_2(x) + ip_3(x) \\ p_2(x) - ip_3(x) & q(x) - p_1(x) \end{pmatrix}.$$

Ansatz for specific data and linear functions q and p are precisely calculated and emphasize a transfer of energy between the two modes at leading order (see also [10] in the case $p_3 = q = 0$, $p_1 = \xi_1$, $p_2 = \xi_2$ and [5] for generic cases with Wigner measure approach).

b) The study of the motion of electrons in a crystal leads to the study of Hamiltonians of the form

$$H(t, x, \xi) = V(t, x) + \begin{pmatrix} q(\xi) + p_1(\xi) & p_2(\xi) + ip_3(\xi) \\ p_2(\xi) - ip_3(\xi) & q(\xi) - p_1(\xi) \end{pmatrix}.$$

In the case $V = V(x)$, $p_1 = \xi_1$, $p_2 = \xi_2$, $q = p_3 = 0$, the crossing has been studied from a Wigner measures point of view in [7].

c) Another example consists in a Dirac type equation (see [4] and [6])

$$H(t, x, \xi) = V(t, x) + \mathcal{P}(\xi - A(t, x)),$$

where

$$\mathcal{P} = \begin{pmatrix} \eta_1 & \eta_2 + i\eta_3 \\ \eta_2 - i\eta_3 & -\eta_1 \end{pmatrix}.$$

The first model studied is Landau and Zener's in the decade 1920 to 1930 (see [16] and [18]),

$$i\varepsilon\partial_t \psi^\varepsilon = \begin{pmatrix} t & x_1 + ix_2 \\ x_1 - ix_2 & -t \end{pmatrix} \psi^\varepsilon.$$

The crossing occurs at $t = x_1 = x_2 = 0$. Both critical trajectories lie above the line $\{x_1 = x_2 = 0\}$. Stretching variables

$$t = \sqrt{\varepsilon} s, \quad x = \frac{y}{\sqrt{\varepsilon}}$$

leads to a scattering problem for the system

$$i\partial_s \psi = \begin{pmatrix} s & y_1 + iy_2 \\ y_1 - iy_2 & -s \end{pmatrix} \psi.$$

It is possible to calculate explicitly the scattering matrix in terms of y. This yields in particular the transmission coefficient for the Wigner measure (or just $\psi \otimes \bar\psi$):

$$T(y) = e^{-\pi|y|^2},$$

$$|y| = \frac{\text{distance to critical trajectories}}{\sqrt{\varepsilon}}.$$

This quick analysis of the Landau and Zener problem emphasizes a point which will be crucial to analyze crossings in more general settings: the energy transfer depends on a second scale $\sqrt{\varepsilon}$. In the following, we shall introduce an enlarged phase space including rescaled coordinates, normal to the set of critical trajectories. With this generalized phase space, we will associate some new Wigner measures, more accurate than the usual ones. This analysis will allow us to prove the existence of a universal (i.e., independent of the data) transmission coefficient for these measures and compute it in terms of the geometry of the system.

In [2] and [3], Y. Colin de Verdière has proved recently a very refined normal-form result in the spirit of the work of [1] for symmetric systems. In view of the invariance of Wigner measures through canonical transforms, it is likely that his result implies ours.

As a conclusion to this introduction, let us mention that an expanded version of this note will appear in [9]. The case of symmetric systems is described in [8].

Summary:

1. Two-scale Wigner measures.
2. Geometric assumptions on the system.
3. Statement of the result.
4. Main steps of the proof.
5. Concluding remarks.

5.2 Two-scale Wigner measures

We consider a classical phase space $T^*(I\!\!R^D)$ and $f = (f_1, \ldots, f_m)$, m smooth and independent functions on $T^*(I\!\!R^D)$ satisfying the commutating relations

$$\{f_j, f_k\} = 0, \quad 1 \leq j, k \leq m.$$

Then, $J = \{f = 0\}$ is an involutive submanifold of $T^*(I\!\!R^D)$.

We define a new class $\mathcal{A}$ of testing symbols

$$a = a(z, \zeta, \eta) \in \mathcal{C}^\infty(I\!\!R_z^D \times I\!\!R_\zeta^D \times I\!\!R_\eta^m),$$

compactly supported in (z, ζ) *with radial limits at infinity in* η,

$$a(z, \zeta, R\eta) \to a_\infty\left(z, \zeta, \frac{\eta}{|\eta|}\right), \quad R \to \infty.$$

Therefore, a extends to a function on $I\!\!R_z^D \times I\!\!R_\zeta^D \times \overline{I\!\!R}_\eta^m$ where $\overline{I\!\!R}^m$ is the compactification of $I\!\!R^m$ with a sphere at infinity.

Theorem 1. *Let (ψ^ε) be any bounded family in $L^2(I\!\!R^D, C^N)$. Up to extracting a subsequence, there exists a (matrix-valued) positive measure ν_f on $J \times \overline{I\!\!R}^m$ such that, for every $a \in \mathcal{A}$,*

$$\int_{T^*(I\!\!R^D)} W\psi^\varepsilon(z, \zeta)\, a\left(z, \zeta, \frac{f(z, \zeta)}{\sqrt{\varepsilon}}\right)\, \mathrm{d}z\, \mathrm{d}\zeta$$

$$\xrightarrow[\varepsilon \to 0]{} \int_{J \times \overline{I\!\!R}^m} a(z, \zeta, \eta)\, \mathrm{d}\nu_f(z, \zeta, \eta) + \int_{T^*(I\!\!R^D)\backslash J} a_\infty\left(z, \zeta, \frac{f(z, \zeta)}{|f(z, \zeta)|}\right)\, \mathrm{d}\mu(z, \zeta)$$

where μ is a Wigner measure of ψ^ε.

- Notice that the measure $\mathbf{1}_J\,\mu$ can be recovered from ν_f through

$$\mathbf{1}_J\,\mu(z,\zeta) = \int_{\overline{\mathbb{R}^m}} \nu_f(z,\zeta,\mathrm{d}\eta)\,.$$

- **Intrinsic definition of ν.** If $g = M \cdot f$ is another system of commuting equations of J,

$$\int_{J\times\overline{\mathbb{R}^m}} a(z,\zeta,\eta')\,\mathrm{d}\nu_g(z,\zeta,\eta') = \int_{J\times\overline{\mathbb{R}^m}} a(z,\zeta,M\cdot\eta)\,\mathrm{d}\nu_f(z,\zeta,\eta).$$

Hence one can define a measure ν_J on $\overline{N}(J)$, the compactified normal bundle above J which is the bundle above J with fibres obtained by adding a sphere at infinity to those of $T(T^*(\mathbb{R}^D))/T(J)$. Then, for every system f of commuting equations of J, we have

$$\int_{\overline{N}(J)} a(\rho,\mathrm{d}f(\rho)\cdot[\delta\rho])\,\mathrm{d}\nu_J(\rho,[\delta\rho]) = \int_{J\times\overline{\mathbb{R}^m}} a(z,\zeta,\eta)\,\mathrm{d}\nu_f(z,\zeta,\eta).$$

- **Localization and propagation of ν_J.** Assume ψ^ε is a solution of a scalar equation

$$\mathrm{op}^\varepsilon(H)\psi^\varepsilon = 0$$

and J is involutive and contained in the energy surface $\Sigma = \{H = 0\}$. Then

  - ν_J is supported in the compactification of $T(\Sigma)/T(J)$.
  - ν_J is invariant by the linearization of the Hamiltonian flow of H. Indeed,

if

$$\phi_s : \Sigma \to \Sigma, \quad \phi_s(J) \subset J$$

denotes the Hamiltonian flow, the linearized flow

$$D\phi_s : T(\Sigma)/T(J) \to T(\Sigma)/T(J),$$

can be extended to the spherical compactifications.

5.3 Geometric assumptions on the system

Consider the Hamiltonian

$$H = k\,\mathrm{Id} + \begin{pmatrix} p_1 & p_2 - ip_3 \\ p_2 + ip_3 & -p_1 \end{pmatrix}.$$

The energy surface is

$$\Sigma = \{(\tau + k)^2 = |p|^2\} \subset T^*(\mathbb{R} \times \mathbb{R}^d)_{((t,x),(\tau,\xi))},$$

and the eigenvalues are

$$E_{\pm} = k \pm |p|.$$

Assume p_1, p_2, p_3 are independent, and let

$$S = \{p_1 = p_2 = p_3 = \tau + k = 0\}.$$

Introduce the electric and magnetic fields as vector-valued functions on the space-time phase space

$$E = \{\tau + k, p\}, \quad B = (\{p_3, p_2\}, \{p_1, p_3\}, \{p_2, p_2\}).$$

Remark that if $k = V(t, x)$ and $p = \xi - A(t, x)$, then $E = -\nabla_x V - \partial_t A$ and $B = \nabla \times A$.

We need that the Hamiltonian curves of $\lambda_{\pm} = \tau + E_{\pm}$ are transversal to S and can be continued through S. This corresponds to one of the following two situations on S,

$$
\begin{aligned}
(H1) \quad & E \cdot B = 0, \ |E|^2 > |B|^2, \\
(H2) \quad & E \cdot B \neq 0.
\end{aligned}
$$

We assume $(H1)$ identically on S, which covers the cases of the Born–Oppenheimer approximation and of the motion in a crystal, where $B = 0$. Then define $J^{\pm,\mathrm{in}}$ as the union of (incoming) trajectories for $\lambda_{\pm}$ arriving on S and $J^{\pm,\mathrm{out}}$ the union of (outgoing) trajectories for $\lambda_{\pm}$ leaving from S. One can prove that these four manifolds are involutive, of codimension 3, and that $J^{\pm,\mathrm{in}}$ connects smoothly to $J^{\mp,\mathrm{out}}$. We can therefore define the four corresponding two-scale Wigner measures

$$\nu_{J^{\pm,\mathrm{in}}}, \quad \nu_{J^{\pm,\mathrm{out}}}.$$

These measures are transported by the linearized Hamiltonian flows of $\tau + E_{\pm}$, which are transversal to S. Therefore, they have traces on S, denoted by

$$\nu_S^{\pm,\mathrm{in}}, \quad \nu_S^{\pm,\mathrm{out}}$$

which are measures on bundles above S with fibres obtained by compactifications of the planes $T(\Sigma^{\pm})/T(J^{\pm})$.

5.4 Statement of the result

The planes $T(\Sigma^{\pm})/T(J^{\pm})$ can be identified to the normal plane of the electric field E through the map

$$[\delta\rho] \mapsto Y = \mathrm{d}(\tau + k) \cdot \delta\rho \, B + \mathrm{d}p \cdot \delta\rho \times E \, .$$

Theorem 2. *Assume* $\nu_S^{-,in} = 0$. *Then*

$$\nu_S^{-,out} = T\nu_S^{+,in}\ , \quad \nu_S^{+,out} = (1-T)\nu_S^{+,in}$$

with

$$T = \exp\left[-\frac{\pi}{(|E|^2 - |B|^2)^{3/2}}\left(|Y|^2 - \left(\frac{E \times B}{|E|^2}\cdot Y\right)^2\right)\right]$$

Notice in particular that:

• On the circle at infinity, $T = 0$: there is total reflection on the $+$ mode.

• If $B = 0$ and $|E| \gg 1$, then $T \to 1$ on the set $|Y| < +\infty$. In other words, the mode conversion is total for high electric fields (in the same spirit, see [6] for analysis of Dirac equations with high electromagnetic fields).

5.5 Main steps of the proof

Step 1. First (rough) normal form and analysis on the circle at infinity.

Near $\rho_0 \in S$, there exists a change of canonical coordinates

$$\kappa : (t, x, \tau, \xi) \mapsto (s, z, \sigma, \zeta)$$

and an invertible matrix $A = A(t, x, \tau, \xi)$ such that, if U is a unitary Fourier integral operator associated to κ, the new unknown

$$v^\varepsilon = U\,\mathrm{op}_\varepsilon(A)\psi^\varepsilon$$

satisfies

$$\mathrm{op}_\varepsilon\begin{pmatrix} \sigma + s & \gamma_1\zeta_1 + \gamma_2\zeta_2 \\ \overline{\gamma}_1\zeta_1 + \overline{\gamma}_2\zeta_2 & \sigma - s \end{pmatrix} v^\varepsilon \sim 0$$

where $\gamma_j = \gamma_j(s, z, \sigma, \zeta)$, $Im(\gamma_1\overline{\gamma}_2) \neq 0$.

Moreover, in these new coordinates,

$$\begin{aligned}
\Sigma &= \{\sigma^2 = s^2 + \zeta_1^2 + \zeta_2^2\}, \\
S &= \{s = \sigma = \zeta_1 = \zeta_2 = 0\}, \\
J^{\pm,in} &= \{\sigma \mp s = \zeta_1 = \zeta_2 = 0, s < 0\}, \\
J^{\pm,out} &= \{\sigma \pm s = \zeta_1 = \zeta_2 = 0, s > 0\}.
\end{aligned}$$

By pseudodifferential energy estimates, one can prove the reflection of the energy located at infinity, i.e., for large scales of oscillations with respect to $\sqrt{\varepsilon}$, $\{(\zeta_1^2 + \zeta_2^2)^{1/2} \gg \sqrt{\varepsilon}\}$; in other words,

$$\nu_S^{\pm,in} = \nu_S^{\pm,out} \quad \text{on the circle at infinity.}$$

Step 2. Analysis at finite distance.

Using that $|\zeta_j| \sim \sqrt{\varepsilon}$, one can refine the normal form and assume that $\gamma_j = \gamma_j(z, \zeta)$. We are reduced to the system

$$i\varepsilon\partial_s u^\varepsilon = \begin{pmatrix} s\,I & \sqrt{\varepsilon}\,G \\ \sqrt{\varepsilon}\,G^* & -s\,I \end{pmatrix} u^\varepsilon$$

where

$$G = \mathrm{op}_\varepsilon\left(\left(\gamma_1\frac{\zeta_1}{\sqrt{\varepsilon}} + \gamma_2\frac{\zeta_2}{\sqrt{\varepsilon}}\right) \chi\left(\frac{\zeta_1}{R\sqrt{\varepsilon}}, \frac{\zeta_2}{R\sqrt{\varepsilon}}\right)\right)$$

and χ is a cutoff function with R large enough. One can then conclude using the following operator-valued Landau–Zener formula.

Theorem 3. *Let $G = G^\varepsilon$ be a bounded family of operators on a Hilbert space $\mathcal{H}$. Let $u^\varepsilon \in C(\mathbb{R}_s, \mathcal{H} \times \mathcal{H})$ be a solution of*

$$i\varepsilon\partial_s u^\varepsilon = \begin{pmatrix} s\,I & \sqrt{\varepsilon}\,G \\ \sqrt{\varepsilon}\,G^* & -s\,I \end{pmatrix} u^\varepsilon$$

with $\sup_{\varepsilon,s} |u^\varepsilon|_\mathcal{H} < +\infty$.

There exist families of vectors, $(\alpha_j^\varepsilon), (\omega_j^\varepsilon)$, $j = 1, 2$, such that for any cutoff function $\chi \in C_0^\infty(\mathbb{R})$, $\chi(GG^*)\alpha_1^\varepsilon$, $\chi(G^*G)\alpha_2^\varepsilon$, $\chi(GG^*)\omega_1^\varepsilon$ and $\chi(G^*G)\omega_2^\varepsilon$ are bounded in $\mathcal{H}$ and such that up to a small error in $\mathcal{H}$,

$$\chi(GG^*)u_1^\varepsilon(s) \sim \chi(GG^*)e^{-is^2/2\varepsilon}\left|\frac{s}{\sqrt{\varepsilon}}\right|^{-iGG^*/2} \alpha_1^\varepsilon \,, \; s < 0\,,$$

$$\chi(GG^*)u_1^\varepsilon(s) \sim \chi(GG^*)e^{-is^2/2\varepsilon}\left|\frac{s}{\sqrt{\varepsilon}}\right|^{-iGG^*/2} \omega_1^\varepsilon \,, \; s > 0\,,$$

$$\chi(G^*G)u_2^\varepsilon(s) \sim \; \chi(G^*G)e^{is^2/2\varepsilon}\left|\frac{s}{\sqrt{\varepsilon}}\right|^{iG^*G/2} \alpha_2^\varepsilon \,, \; s < 0\,,$$

$$\chi(G^*G)u_2^\varepsilon(s) \sim \; \chi(G^*G)e^{is^2/2\varepsilon}\left|\frac{s}{\sqrt{\varepsilon}}\right|^{iG^*G/2} \omega_2^\varepsilon \,, \; s > 0\,.$$

Moreover

$$\begin{pmatrix} \omega_1^\varepsilon \\ \omega_2^\varepsilon \end{pmatrix} = \begin{pmatrix} a(GG^*) & -b(GG^*)G \\ \bar{b}(G^*G)G^* & a(G^*G) \end{pmatrix} \begin{pmatrix} \alpha_1^\varepsilon \\ \alpha_2^\varepsilon \end{pmatrix},$$

with

$$a(\lambda) = e^{-\pi\lambda/2},$$

$$b(\lambda) = 2ie^{i\pi/4}\lambda^{-1}\pi^{-1/2}2^{-i\lambda/2}e^{-\pi\lambda/4}\Gamma\left(1 + i\frac{\lambda}{2}\right)\sinh(\pi\lambda/2),$$

$$a(\lambda)^2 + \lambda|b(\lambda)|^2 = 1.$$

5.6 Concluding remarks: Advantages and disadvantages of Wigner measures in the analysis of eigenvalue crossings

• Wigner measures lead to explicit and geometric formulae for leading order terms of Wigner transforms, independently of the specific expression of the initial wave functions. Moreover, they allow a global and fairly simple description of the leading dynamics (see, e.g., [10] for an example in the Born–Oppenheimer context).

• On the other hand, this analysis, because of its focusing on quadratic objects, does not provide information about possible interferences between the two modes. This information would necessitate a more precise description of the wave function (as in the last theorem). The corresponding geometric objects (concentration profiles on the involutive manifolds J?) are still to be introduced for arbitrary data. Moreover, it would be interesting to have a more accurate description including lower order terms.

References

1. P.J. Braam, J.J. Duistermaat, Normal forms of real symmetric systems with multiplicity, *Indag. Mathem.* N.S., **4** (4) (1993), 407–421.
2. Y. Colin de Verdière, The level crossing problem in semi-classical analysis I. The symmetric case, *Ann. Inst. Fourier* **53** (2003), 1023–1054.
3. Y. Colin de Verdière, The level crossing problem in semi-classical analysis II. The hermitian case, *Preprint de l'Institut Fourier*, 2003.
4. C. Fermanian Kammerer, A non-commutative Landau–Zener formula. *Prépublication de l'Université de Cergy-Pontoise*, 2002, to appear in *Math. Nach.*
5. C. Fermanian Kammerer, Wigner measures and Molecular Propagation through Generic Energy Level Crossings, *Prépublication de l'Université de Cergy-Pontoise*, October 2002.
6. C. Fermanian Kammerer, Semi-classical analysis of a Dirac equation without adaibatic decoupling, *Prépublication de l'Université de Cergy-Pontoise*, 2003; to appear in *Monat. Für Math.*
7. C. Fermanian Kammerer, P. Gérard, Mesures semi-classiques et croisements de modes, *Bull. Soc. Math. France*, **130** $N^o 1$, (2002), 123–168.
8. C. Fermanian Kammerer, P. Gérard, Une formule de Landau–Zener pour un croisement générique de codimension 2, *Séminaire E.D.P.* 2001–2002, Exposé $N^o 21$, Ecole Polytechnique.
   http://math.polytechnique.fr/seminaires/seminaires-edp/
9. C. Fermanian Kammerer, P. Gérard, A Landau–Zener formula for non-degenerated involutive codimension 3 crossing, *Prépublication de l'Université de Cergy-Pontoise*, Sept. 2002; *Annales Henri Poincaré* **4** (2003), 513–552.
10. C. Fermanian Kammerer, C. Lasser, Wigner measures and codimension two crossings, *Jour. Math. Phys.* **44** 2 (2003), 507–527.
11. P. Gérard, P. A. Markowich, N. J. Mauser, F. Poupaud, Homogenization Limits and Wigner Transforms, *Comm. Pure Appl. Math.*, **50** (1997), 4, 323–379 and **53** (2000), 280–281.

12. G.A. Hagedorn, Proof of the Landau–Zener formula in an adiabatic limit with small eigenvalue gaps, *Commun. Math. Phys.* **136** (1991), 433–449.

13. G.A. Hagedorn, Molecular Propagation through Electron Energy Level Crossings, *Memoirs of the A. M. S.*, **111** N° 536, (1994).

14. G.A. Hagedorn, A. Joye, Landau–Zener transitions through small electronic eigenvalue gaps in the Born–Oppenheimer approximation, *Ann. Inst. Henri Poincaré, Physique Théorique*, **68** N°1, (1998), 85–134.

15. G. A. Hagedorn, A. Joye, Molecular propagation through small avoided crossings of electron energy levels, *Rev. Math. Phys.* **11** N° 1, (1999), 41–101.

16. L. Landau: *Collected Papers of L. Landau*, Pergamon Press, 1965.

17. P-L. Lions, T. Paul: Sur les mesures de Wigner, *Revista Matemática Iberoamericana* **9** (1993), 553–618.

18. C. Zener: Non-adiabatic crossing of energy levels, *Proc. Roy. Soc. Lond.* **137** (1932), 696–702.

6

Stability of Three- and Four-Body Coulomb Systems

A. Martin

Theoretical Physics Division, CERN
CH - 1211 Geneva 23
and
LAPP - F 74941 ANNECY LE VIEUX

Summary. We discuss the stability of three- and four-particle systems interacting by pure Coulomb interactions, as a function of the masses and charges of the particles. We present a certain number of general properties which allow us to answer a certain number of questions without or with fewer numerical calculations.

6.1 Introduction

In this talk, I would like to speak of the problem of the stability of three- and four-body non-relativistic purely Coulombic systems. A system will be said to be stable if its energy is lower than the energy of any subdivision in subsystems. This is a restrictive definition of stability, because besides that there are other useful notions: "metastability" and "quasi-stability" on which we shall say only a few words later. The reason I chose this subject when I was invited to the conference on multiscale methods in quantum mechanics is that in the work, scaling is used a lot and in different ways: various scalings of the masses, scaling of the charges. The works I will present are due, in what concerns the three-body case, to J.-M. Richard, T.T. Wu and myself. The four-body work is due to J.-M. Richard in collaboration with various other persons, including J. Fröhlich, a participant in this conference. I shall speak of:

i) three-body systems with equal absolute value of the charge, i.e., $-e+e+e$, or $+e-e-e$, since it is clear that $+e+e+e$ is unbound. Then binding or no binding will depend on the masses;
ii) three-body systems with unequal charges;
iii) four-body systems with charges with equal absolute value. It will be mostly $+e+e-e-e$. However, I shall say a word on $+e-e-e-e$.

6.2 Three-body case: Equal |charges|

The problem we discuss now is whether a system of three charged particles (1,2,3), 1 having charge $+e$ and 2 and 3 charges $-e$, is stable or will dissociate into a two-body system and an isolated particle, (1,2)+3 or (1,3)+2. The system will be stable if the algebraic binding energy of the (1,2,3) system is strictly less than the binding energy of both (1,2) and (1,3). If, on the other hand, the infimum of the spectrum of the (1,2,3) system coincides with the lowest of the (1,2) and (1,3) binding energies the system will be unstable. This is an old problem which has been treated in many particular cases. For instance, long ago, Bethe has shown that the hydrogen negative ion (pe^-e^-) has one bound state [4], and Hill has shown that there is only one such bound state with natural parity [12], and Drake has also shown that there exists an unnatural parity state [6] and finally Grosse and Pittner [11] have shown also that this unnatural parity state is unique. In what follows we shall treat only the natural parity states, i.e., states such that $P = (-1)^L$, where L is the total orbital angular momentum (we neglect spin interactions!). For three particles there is no problem with the Pauli principle even if two of them are identical fermions, since we can adjust the spin. Wheeler [24] has also shown that the system $e^+e^-e^-$ is bound, and, more generally, Hill [12] has shown that any three-body system in which the two particles with the same sign of the charge have the same mass is stable. This covers the two previous cases. As an example of an unstable system (there are many others !), we can give the proton-electron-negative muon system, for which a heuristic proof was given by Wightman in his thesis [25] and a rigorous proof was given by a collaboration including Glaser, Grosse, Thirring and I [10]. Richard, Wu and I [17] have tried to organise the results on stability, and, by using simple properties, save numerical calculations. From the reactions we had from experts on numerical calculations we believe that this was not totally useless. The three-body Schrödinger equation reads

$$-\frac{1}{2m_1}\Delta_1\psi - \frac{1}{2m_2}\Delta_2\psi - \frac{1}{2m_3}\Delta_3\psi + \left[-\frac{e^2}{r_{12}} - \frac{e^2}{r_{13}} + \frac{e^2}{r_{23}}\right]\psi = E\psi \quad (6.1)$$

and the corresponding two-body equations can be obtained by omitting some terms. It is obvious that we have *scaling* properties:

i)   the charges can be multiplied by some arbitrary number without changing the stability problem;

ii) the masses can also be multiplied by an arbitrary number, so that the stability problem depends only on the *ratio* of the masses, i.e., of *two* parameters.

It will be convenient to introduce some variables:

* the inverse of the masses

$$x_1 = \frac{1}{m_1} \quad x_2 = \frac{1}{m_2} \quad x_3 = \frac{1}{m_3}, \qquad (6.2)$$

then the ground state energy of the system will be concave in x_1, x_2, x_3, and, in particular, concave in x_1 when x_2 and x_3 are fixed (and circular permutations!);

- the *constrained* inverse of the masses

$$\alpha_1 = \frac{x_1}{x_1 + x_2 + x_3} \quad \text{etc.} \qquad (6.3)$$

such that

$$\alpha_1 + \alpha_2 + \alpha_3 = 1 . \qquad (6.4)$$

With these new variables, any system of three particles can be represented by a point in a triangle, $\alpha_1, \alpha_2, \alpha_3$ being the distances to the sides of the triangle. Figure 6.1 represents such a triangle with a few points representing some three-body systems.

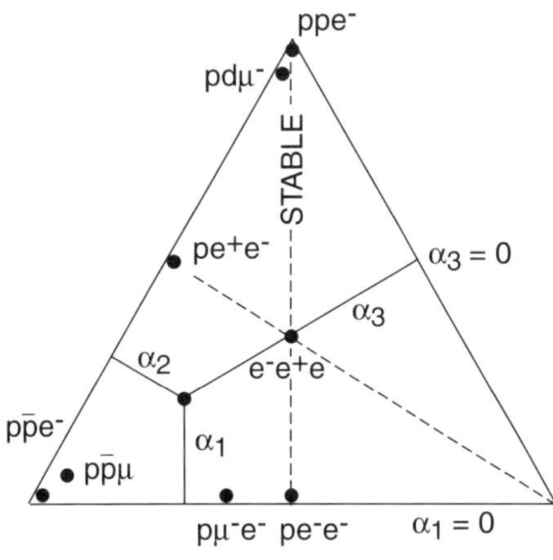

Fig. 6.1.

It is of course sufficient, for the time being (i.e., for *equal* charges of 2 and 3) to consider the left half of the triangle, i.e., to assume $m_2 \geq m_3$. Let us remember that since we have, according to Hill's theorem, strict stability for $m_2 = m_3$, i.e., $\alpha_2 = \alpha_3$, there will be some neighbourhood of the line $\alpha_2 = \alpha_3$ where we shall have stability. However, not all systems will be stable. We have already mentioned the $pe^-\mu^-$ system as unstable. Another point where instability is obvious is the left summit marked 3, where we have two infinitely heavy particles with opposite charge producing zero attraction on the third particle. There is, therefore, an instability region in the left-half triangle. We have proved three theorems on the instability region in the left-half triangle.

Theorem 1. *The instability region in the left-half triangle is star-shaped with respect to summit 3.*

The proof is based on the Feynman-Hellmann theorem combined with scaling. take a point P (Fig. 6.2) where the system is unstable or at the limit of stability. First we use the variables x_1, x_2, x_3. From the Feynman–Hellmann theorem, $\frac{dE(123)}{dx_3} > 0$, if x_1 and x_2 are fixed. The binding energy of the subsystem 12 is fixed. Hence the residual binding can only increase (algebraically). x_3 moves from $x_3(P)$ to infinity. The image of this in the rescaled α variable is the segment, $P3$, where $\alpha_1/\alpha_2 = $ constant. If there is no binding at P there is no binding on the whole segment.

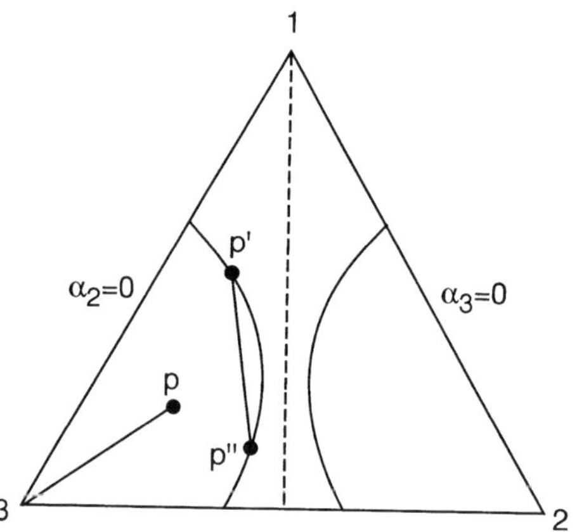

Fig. 6.2.

Theorem 2. *In the left-half triangle, the instability region is* convex.

Take two points P' and P'' on the border of the stability domain inside the triangle with the α variables. At P' and P'' we have $E_{P'}(12) = E_{P'}(123)$, $E_{P''}(12) = E_{P''}(123)$. It is possible to find a linear rescaling $P \to M$ such that $E_{M'}(12) = E_{M''}(12) = E_{M'}(123) = E_{M''}(123)$. Then one can interpolate linearly between M' and M'':

$$M_\lambda = \lambda M' + (1 - \lambda)M'' , \qquad 0 < \lambda < 1 \ .$$

For any M_λ, $E_{M_\lambda}(12) = $ const. and $E_{M_\lambda}(123)$ is *concave* in λ, and therefore $E_{M_\lambda}(123) \geq E_{M'}(123) = E_{M''}(123)$. Returning to the original variables $\alpha_1\alpha_2\alpha_3$ and noticing that the scaling is *linear* we see that, on $P'P''$ we have $E(123) \geq E(12)$. Hence we have instability (Fig. 6.2). There is, in fact, a more refined theorem, which we found, following a question by the late V.N. Gribov during a seminar in Budapest in 1996.

Theorem 3. *The domain (in the left-half triangle) where*

$$\frac{E(123)}{E(12)} \le 1 + \epsilon \ , \qquad \epsilon > 0$$

is convex.

The meaning of this theorem is that the lines along which the *relative* binding is constant have a definite convexity. Note the *sign* of the inequality because $E(123)$ and $E(12)$ are both *negative*. We believe that the proof is essentially obvious, since, in the previous theorem, one goes through a rescaling, replacing P and P' by M and M' where the two-body energies are equal. Theorem 2 is of course becoming a special case of Theorem 3, with $\epsilon \to 0$. Let us give a very simple application of Theorem 2. We know that, according to Glaser et al., the system $p_\infty A^- e^-$ is unstable if $m_{A^-} > 1.57 m_{e^-}$ (p_∞ means a proton with infinite mass). Similarly, we know, from the work of Armour and Schrader [1], that the system $p_\infty A^+ B^-$ is unstable if $m_{A^+}/m_{B^-} < 1.51$. This means that $p_\infty e^- e^+$ is unstable (not because of annihilation that we neglect, but of dissociation into $p_\infty e^-$ and e^+). In Fig. 6.3, $p_\infty A^+ B^-$ and $p_\infty A^- e^-$ with the limit masses corresponding respectively to X and Y. Any point to the left of the segment XY corresponds, according to Theorem 2, to an unstable system. Therefore, $p e^+ e^-$, $p \mu^+ \mu^-$, with the *actual mass* of the proton, are unstable, and one can go up to $p z^- z^+$ which will be unstable if $m_p/m_z > 2.2$.

We obtain too that the system $p \mu^- e^-$, with the actual proton mass is unstable, and also (disregarding again annihilation) $p\bar{p} e^-$, $p\bar{p} \mu^-$. One can also use convexity to get results in the opposite direction, i.e., prove that certain three-body systems are stable. We know that the systems represented by a point on the vertical bisector of the triangle are stable. In practice we know more than that, namely we have an estimate or more exactly a lower bound of the absolute value of the binding energy of many systems by using variational calculations and, in fact, by playing with convexity again it is possible to have a lower bound of the absolute value of the binding energy at *any* point on the bisector which corresponds to $\alpha_2 = \alpha_3$, $0 \le \alpha_2 \le 1$. Now we use convexity along a horizontal line $\alpha_1 = \text{const}$. The systems $\alpha_1, \alpha_2, \alpha_3$, $\ \alpha_1, \alpha_3, \alpha_2$ represented in Fig. 6.2 by Q and Q', are of course completely equivalent. Hence

$$E_{123}(\alpha_1, \alpha_2, \alpha_3) = E_{123}(\alpha_1, \alpha_3, \alpha_2) < E_{123}\left(\alpha_1, \frac{\alpha_2 + \alpha_3}{2}, \frac{\alpha_2 + \alpha_3}{2}\right) \ ,$$

by convexity and

$$E_{123}\left(\alpha_1, \frac{\alpha_2 + \alpha_3}{2}, \frac{\alpha_2 + \alpha_3}{2}\right) = \left(1 + g(\alpha_1)\right) E_{12}\left(\alpha_1, \frac{\alpha_2 + \alpha_3}{2}\right)$$
$$= \left(1 + g(\alpha_1)\right) E_{12}\left(\alpha_1, \frac{1 - \alpha_1}{2}\right) \ ,$$

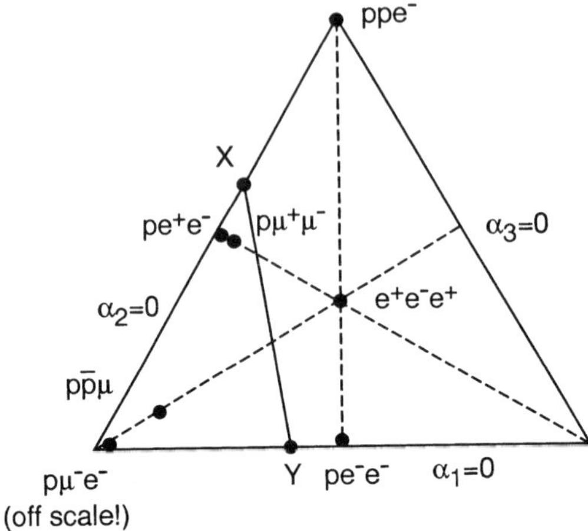

Fig. 6.3.

where g represents the relative excess in binding energy. We are assured of stability if

$$E_{12}(\alpha_1, \alpha_2) > \left(1 + g(\alpha_1)\right) E_{12}\left(\alpha_1, \frac{1 - \alpha_1}{2}\right),$$

i.e., if

$$\frac{e^2}{2} \; \frac{1}{\alpha_1 + \alpha_2} < \left(1 + y(\alpha_1)\right) \frac{e^2}{2} \; \frac{2}{1 + \alpha_1}.$$

In this way, it is possible to prove that the system $pd\mu^-$, important for fusion processes, is stable, though it is off the diagonal. One would like to show also in this way that $\pi^+\mu^-\mu^+$ and $\mu^+\pi^+\pi^-$ are stable, but these considerations are not sufficient. There is a hint that they are stable because, from explicit calculations at $\alpha_1 = 0$, one sees that this method tends to give a band of stability which is two times narrower than the real one and this is just what one needs.

6.3 Three-body case. Unequal charges

On this topic, Richard, Wu and I have published one paper [18]. We have the right to take $q_1 = 1$, the charge of the particle which is opposite to the other two, of the same sign, q_2 and q_3.

A) Unequal charges, but $q_2 = q_3$

This is the simplest case, very similar to the case of all equal charges. For fixed $q_2 = q_3$ we can again represent a system with the variables $\alpha_1\alpha_2\alpha_3$, and, on the bisector of summit 1, the energies of the subsystems (12) and (13) are equal. The fact that the instability regions are star-shaped with respect to 3 for the left-half of the triangle and to 2 for the right-half persists, and as well the convexity of the instability regions. There are two major differences which are:

i) that if $q_2 = q_3 < 1$, all three-body systems are stable, because near summit 3, for instance, the subsystem (12) is very compact and exerts a Coulomb attraction at long distances on particle 3; it may seem strange that as $q_2 \to 1$ part of the triangle becomes unstable, but this is just due to the fact that the binding energy, in that region, tends to zero as $q_2 \to 1$;

ii) that if $q_2 = q_3$ is large enough, stability disappears completely.

Figure 6.4 summarizes the situation. For $q_2 > 1$ but very close to 1, there is no qualitative difference, but for a certain critical value $1 < q_{2c} < 1.1$, the stability band breaks into two pieces, and from calculations by Hill and collaborators [2] stability near $\alpha_2 = \alpha_3 = 1/2$, $\alpha_1 = 0$ disappears completely for $q_2 \geq 1.1$, and from the calculations of Hogrève it disappears near $\alpha_2 = \alpha_3 = 0$, $\alpha_1 = 1$ for $q_2 > 1.24$. From convexity, it hence disappears *completely* along the segment joining $\alpha_2 = \alpha_3 = 0$ and $\alpha_2 = \alpha_3 = 1/2$, and from the star-shaped property, there is no stability at any point in the triangle for $q_2 > 1.24$.

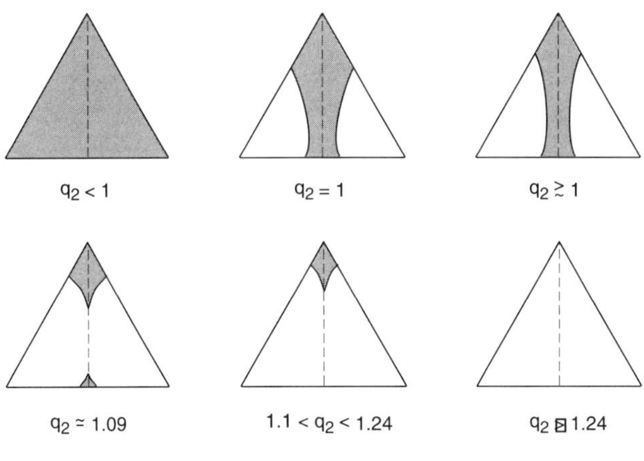

Fig. 6.4.

B) Unequal charges, but $q_2 \neq q_3$ fixed

First we continue to use the $\alpha_1, \alpha_2, \alpha_3$ variables to describe the three-body system for fixed charges. A fundamental difference is that the bisector of summit 1 of the triangle no longer plays a special role. It is, instead, the line along which $E_{12} = E_{13}$, i.e.,

$$\frac{q_2^2}{\alpha_1 + \alpha_2} = \frac{q_3^2}{\alpha_1 + \alpha_3} \quad \text{or} \quad q_2^2\left(1 - \alpha_2\right) = q_3^2\left(1 - \alpha_3\right),$$

which becomes important. This line goes through the point $\alpha_2 = \alpha_3 = 1$, symmetric to summit 1 with respect to the line $\alpha_1 = 0$ (Fig. 6.5). The line divides the triangle into two subregions. If we decide to take $q_2 \geq q_3$, $E_{12} < E_{13}$ in the left region which contains the summit 1.

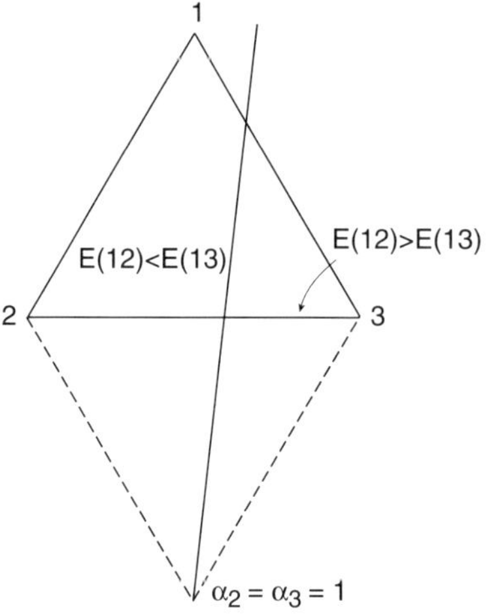

Fig. 6.5.

If q_2 and q_3 are both less than 1, we have again stability everywhere. If $q_2 \geq 1$, part of the triangle becomes unstable. For details, see [15].

C) q_2 and q_3 variable, fixed masses

Instead of holding charges fixed one can fix the masses and study stability in the q_2, q_3 plane.

In the q_2, q_3 plane there is again for the general mass case a dividing line where the binding energies of the two subsystems (12) and (13) are equal:

$$q_2^2 \frac{m_2}{m_1 + m_2} = q_3^2 \frac{m_3}{m_1 + m_3}.$$

In the two sectors thus defined there are two instability regions for which we have been able to derive a new concavity property:

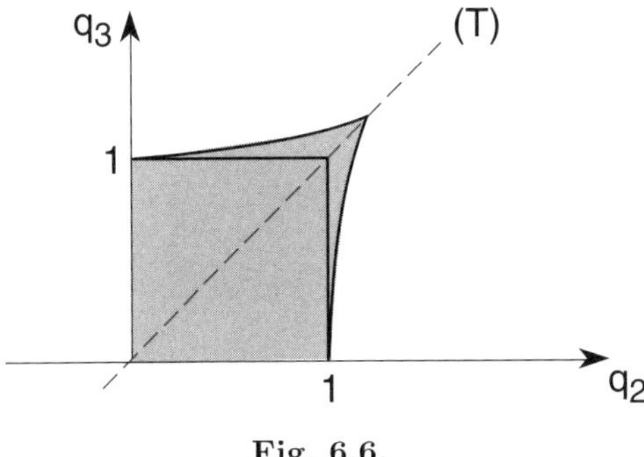

Fig. 6.6.

Theorem 4. *Define $z_2 = 1/q_2, z_3 = 1/q_3$, the image of $q_2 > 0\ q_3 > 0$ is $z_2 > 0\ z_3 > 0$. Then, in the z variables the two instability regions are convex (Fig. 6.7). The proof is based on a rescaling such that the binding energy of the relevant subsystem remains constant on a segment in the $z_2 z_3$ plane.*

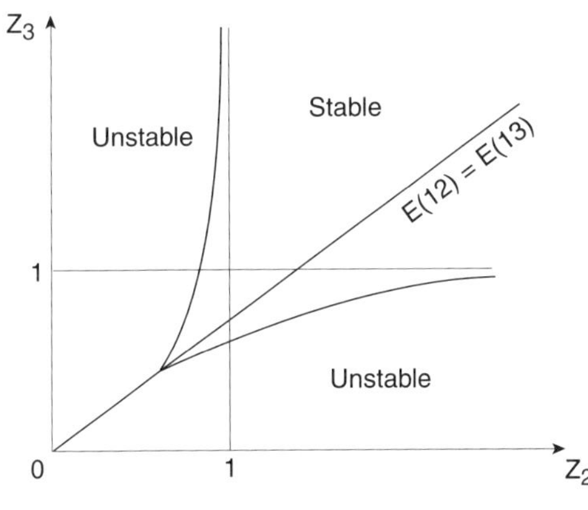

Fig. 6.7.

D) An illustration: The instability of the systems $\alpha p e^-$ or $\alpha p \mu^-$

In the Born–Oppenheimer limit it is known that such systems are unstable [3]. Spruch and collaborators [5] have given arguments which seem to indicate that this might remain true for the actual masses of the protons and of the α particle, but, to our knowledge, a completely rigorous proof does not exist.

STEP 1 Take $m_2 = m_3$. Then if $q_2 = q_3$ we have instability if $q_2 = q_3 \geq 1.24$ and $m_2 = m_3 = 0$ and $m_2 = m_3 = \infty$, and by concavity for any $m_2 = m_3$.

Now consider the segment $q_2 = 1.24$, $0 < q_3 < 1.24$. Along this segment the subsystem with the most negative binding energy is (12) and this energy is constant. There is no stability for $q_2 = q_3 = 1.24$, and no stability for $q_2 = 1.24$, q_3 very close to zero, because then particle 3 is submitted to a very weak force and is therefore very far away most of the time while (12) is overall repulsive for 3. By concavity there is no stability for $q_2 = 1.24$, $0 < q_3 < 1.24$. The same kind of argument applies to $1.24 < q_3 < \infty$, because one has instability for $q_3 \to \infty$ $q_2 > 1$, and one can use concavity in the inverse charge. The conclusion is that if $m_2 = m_3$, one has no stability if either $q_2 \geq 1.24$ or $q_3 \geq 1.24$.

STEP 2 Assume $q_2 \geq q_3$. Then, in the whole sector, $m_2 > m_3$, the lowest two-body threshold is given by the (12) system. Therefore in the α triangle the region $\alpha_2 < \alpha_3$, is completely unstable for $q_2 > 1.24$ by use of the star shaped property. The systems

$$\alpha p e^- \quad \alpha p \mu^- ,$$
$$\alpha d e^- \quad \alpha d \mu^- ,$$
$$\alpha t e^- \quad \alpha t \mu^-$$

satisfy precisely the conditions: $q_2 > 1.24$, $q_3 = 1$, $m_2 > m_3$, and are therefore unstable, in the sense we have given to "instability". *Notice that the proof would fail if the particle with charge 2 was lighter than the particle with charge 1.* However, as pointed out by for instance Gerstein [9], some of the levels of these systems are "quasi stable" in the Born–Oppenheimer approximation in the sense that the minimum of the Born–Oppenheimer potential is below the value it takes for infinite separation between the two nuclei *where one of the limit atomic states is excited and degenerate with the other one.* For instance, in the $\alpha \mu^- d$ system, there exists a quasi-stable state in which, for large separations, the $\alpha \mu^-$ system is in an N=1 state and the $d \mu^-$ system is in an N=2 state. Gerstein [8] went as far as estimating the lifetimes of these quasi-stable states and showed that the lifetime increases drastically when the proton is replaced by a triton. One should also mention metastability, where the Born–Oppenheimer curve has a minimum above zero [15].

6.4 Four-body case: Equal charges

Most of what I will say concerns systems of two positive and two negative charges of absolute value e. However, let me start with the case

$$pe^- e^- e^- ,$$

i.e., a doubly negative hydrogen ion. Such a state according to a review by Hogrève [13] does not seem to exist. In the limit of an infinitely heavy proton, the Lieb bound on n, the number of electrons around a charge Z [16], $n <$

$2Z + 1$, which is a strict inequality, gives $n < 3$. In fact no doubly negative atomic ions seem to exist in nature, while singly negative ions may (like H^-) or may *not* exist (like the case of the rare gases).

We return now to systems with charges $-e - e + e + e$, and first of all $m^- m^- M^+ M^+$, i.e., two negatively charged particles with equal mass and two positively charged particles with equal masses. A familiar example is the hydrogen molecule $e^- e^- p^+ p^+$. A more exotic example is the positronium molecule $e^- e^- e^+ e^+$. It has been realized by Jurg Fröhlich that up to very recently there did not exist any rigorous proof of the stability of the hydrogen molecule. It was believed to be stable because of experiment of course, and of Born–Oppenheimer calculations. Two groups (Fröhlich et al., Richard) investigated this problem and finally joined their efforts to produce a completely rigorous proof [7]. The simplest approach, whose idea comes from J.-M. Richard [20] consists of starting from the work of Øre, which is valid by scaling for a system $A^- A^- A^+ A^+$ [14]. Øre used a very simple variational trial function, of the form

$$\psi = \exp-\tfrac{1}{2}\left(r_{13} + r_{14} + r_{23} + r_{24}\right),$$

$$\cosh\left[\tfrac{\beta}{2}\left(r_{13} - r_{14} - r_{23} + r_{24}\right)\right].$$

Notice that the distances between particles with same charge sign do not appear. All integrals can be carried analytically and it is found that the energy is less than

$$2.0168 \quad E_0(A^+ A^-)\,.$$

The system is therefore stable because it cannot dissociate into $A^+ A^- + A^+ A^+$. It cannot dissociate either in $A^+ A^- A^+ + A^-$, because between these two systems there is a long-distance Coulomb force, producing unavoidably infinitely many bound states. If we take now

$$x_e + x_p = \frac{1}{m_e} + \frac{1}{m_p} = \frac{2}{m_A}\,,$$

we see that the binding energy of $e^- p$ is the same as that of $A^+ A^-$. However,

$$E(x_e, x_e, x_p, x_p) = E(x_p, x_p, x_e, x_e)$$
$$< E\left(\frac{x_p + x_e}{2}, \frac{x_p + x_e}{2}, \frac{x_p + x_e}{2}, \frac{x_p + x_e}{2}\right),$$

by concavity in the inverse masses. So,

$$E(p^+, p^+, e^-, e^-) < E(A^+, A^+, A^-, A^-) < 2E(A^+ A^-)\,.$$

Hence, *ppee* is stable. One can wonder if stability remains if the masses of two particles of the same charge are different, i.e.,

$$A^+ B^+ C^- C^- \ .$$

Then there is still a unique possible dissociation threshold:

$$A^+ C^- + B^+ C^- \ .$$

Øre has predicted explicitly that the system $pe^+e^-e^-$ is *stable* [19], and this has been observed experimentally by Schräder and collaborators [21]. It is also easy, from the upper bound of the energy of $e^+e^+e^-e^-$ to show that the systems $p\,d\,e^-e^-$, $p\,t\,e^-e^-$, $d\,t\,e^-e^-$ are stable. This is implicit in the work of Richard [20], established in the thesis of Seifert [22] and I present here my own version. By concavity we have

$$E(x_A, x_B, x_C, x_C) < E\left(\frac{x_A + x_B}{2}, \frac{x_A + x_B}{2}, x_C, x_C\right)$$

$$< E\left(\frac{x_A + x_B}{4} + \frac{x_C}{2}, \frac{x_A + x_B}{4} + \frac{x_C}{2}, \frac{x_A + x_B}{4} + \frac{x_C}{2}, \frac{x_A + x_B}{4} + \frac{x_C}{2}\right)$$

$$< -2.0168 \ \frac{1}{4} \ \frac{1}{\dfrac{x_A + x_B}{4} + \dfrac{x_C}{2}} \ .$$

If the inequality

$$-2.0168 \frac{1}{\dfrac{x_A + x_B}{4} + \dfrac{x_C}{2}} < -\left(\frac{1}{\dfrac{x_A}{2} + \dfrac{x_C}{2}} + \frac{1}{\dfrac{x_B}{2} + \dfrac{x_C}{2}}\right)$$

is satisfied, the system is stable. If $m_A > m_B > m_C$, one finds that this condition is certainly satisfied if $m_B > 5m_C$. Using the more refined bound [7]

$$E_{A^+ A^+ A^- A^-} < -2.06392 E(A^+ A^+) \ ,$$

which uses a more sophisticated trial function and must be "cleaned" from numerical roundup errors, one gets

$$m_B > 2.45 m_C \ .$$

However, Varga and collaborators [23], using trial functions leading to integrals which can be expressed analytically, and adjusting parameters, have found that one has stability for any m_A and m_B, including the case where one or two of them are less than m_C. By a tedious but feasible exercise, one could, using concavity, transform this calculation, which is unavoidably done for discrete values of the masses, into a very inelegant proof. Let us hope that someone, in the future, will find a still more clever trial function and avoid this.

References

1. E.A.G. Armour and D.M. Schrader, *Can. J. Phys.* **60** (1982), 581.
2. J.D. Baker, D.E. Freund, R.N. Hill and J.D. Morgan, *Phys. Rev.* **A41** (1990), 1247.
3. D.R. Bates and T.R. Carson, *Proc. Roy. Soc.* (London) **A234** (1956), 207.
4. H.A. Bethe, *Z. Phys.* **57** (1929), 815.
5. Z. Chen and L. Spruch, *Phys. Rev.* **A42** (1990), 133; F.H. Gertler, H.B. Snodgrass and L. Spruch, *Phys. Rev.* **172** (1968), 110.
6. G.W.F. Drake, *Phys. Rev. Lett.* **24** (1970), 126.
7. J. Fröhlich, G.M. Graf, J.-M. Richard and M. Seifert, *Phys. Rev. Lett.* **71** (1993), 1332.
8. S.S. Gerstein and V.V. Gusev, *Hyperfine Interactions* **82** (1993), 185.
9. S.S. Gerstein and L.I. Ponomarev, in *Muon Physics* Vol. III, V.W. Hughes and C.S. Wu, eds., Academic Press, New York, 1975.
10. V. Glaser, H. Grosse, A. Martin and W. Thirring, in *Studies in Mathematical Physics*, E.H. Lieb, B. Simon and A.S. Wightman, eds. Princeton University Press, 1976, 169.
11. H. Grosse and H. Pittner, *J. Math. Phys.* **24** (1982), 1142.
12. R.N. Hill, *J. Math. Phys.* **18** (1977), 2316.
13. H. Hogrève, unpublished, review at the International Workshop on Critical Stability of Few Body Quantum Systems, European Center of Theoretical Studies in Nuclear Physics and Related Areas (ECT*), Trento, February 1997; M. Scheller et al., *Science* **270** (1995), 1160.
14. E.A. Hylleras and A. Øre, *Phys. Rev.* **71** (1947), 493.
15. A. Krikeb, A. Martin, J.-M. Richard and T.T. Wu, *Two-Body Problems* **29** (2000), 237.
16. E.H. Lieb, *Phys. Rev.* **A29** (1984), 3018.
17. A. Martin, J.-M. Richard and T.T. Wu, *Phys. Rev.* **A46** (1992), 3697.
18. A. Martin, J.-M. Richard and T.T. Wu, *Phys. Rev.* **A52** (1995), 2557.
19. A. Øre, *Phys. Rev.* **83** (1951), 665.
20. J.-M. Richard, *Phys. Rev.* **A49** (1994), 3573.
21. D.M. Schräder et al., *Phys. Rev. Lett.* **69** (1992), 57.
22. M. Seifert, Thesis, ETH, Zürich (1993).
23. K. Varga, S. Fleck and J.-M. Richard, *Europhysics Lett.* **37** (1997), 183.
24. J.A. Wheeler, *Ann. N.Y. Acad.Sci.* **48** (12946), 219.
25. A.S. Wightman, Thesis, Princeton University (1946).

7

Almost Invariant Subspaces for Quantum Evolutions

G. Nenciu

Department of Theoretical Physics
University of Bucharest
P.O. Box MG 11
RO-76900 Bucharest, Romania
and
Institute of Mathematics of the Romanian Academy
PO Box 1-764
RO-70700 Bucharest, Romania

Summary. In the first part the general framework of almost invariant subspaces for quantum evolutions (which can be viewed as a far-reaching generalization of the standard reduction theory [Ka] which lies at the core of Rellich–Kato theory of analytic perturbations) is reviewed. As examples, in the second part, (non-convergent) expansions leading to almost invariant subspaces are presented in more detail for time-dependent perturbations, as well as for the semi-classical limit.

7.1 Introduction

Let $U_\varepsilon(t, t_0)$ be the unitary evolution as given by the Schrödinger equation

$$i\frac{d}{dt}U_\varepsilon(t, t_0) = H_\varepsilon(t)U_\varepsilon(t, t_0); \quad U_\varepsilon(t_0, t_0) = 1; \quad \varepsilon > 0 \qquad (7.1)$$

in the limit $\varepsilon \to 0$. Since the direct integration of (7.1) is a difficult task we consider here the problem of obtaining information about $U_\varepsilon(t, t_0)$ without actually solving (7.1) in full generality.

The simplest case of such a program is the Rellich–Kato reduction theory when $H_\varepsilon(t)$ does not depend on time and is of the form $H_\varepsilon = H_0 + \varepsilon V$, with V relatively bounded with respect to H_0. As is well known [Ka], in this case if σ_0 is a bounded isolated part of the spectrum of H_0, then for sufficiently small ε, H_ε still has an isolated bounded part of the spectrum, σ_ε, coinciding with σ_0 in the limit $\varepsilon \to 0$ and moreover the corresponding Riesz projection has, for sufficiently small ε, a norm convergent expansion:

$$P_\varepsilon = \sum_{j=0}^{\infty} \varepsilon^j \, E_j; \quad E_j = (-1)^j \, \varepsilon^j \, \frac{i}{2\pi} \oint_\Gamma (H_0 - z)^{-1}(V(H_0 - z)^{-1})^j dz \quad (7.2)$$

and Γ is a contour enclosing σ_0.

By Stone's theorem, $U_\varepsilon(t, t_0) = e^{-i(t-t_0)H_\varepsilon}$ has a block-diagonal structure with respect to the decomposition given by P_ε: $(1 - P_\varepsilon)U_\varepsilon(t, t_0)P_\varepsilon = 0$ or alternatively $P_\varepsilon = U_\varepsilon(t, t_0)P_\varepsilon U_\varepsilon(t, t_0)^*$ i.e., $P_\varepsilon \mathcal{H}$ is an invariant subspace for the evolution $U_\varepsilon(t, t_0)$. Moreover, if N_ε is the Sz-Nagy matrix intertwining P_ε and P_0,

$$N_\varepsilon = (1 - (P_\varepsilon - P_0)^2)^{-1/2}(P_\varepsilon P_0 + (1 - P_\varepsilon)(1 - P_0)), \quad (7.3)$$

then $N_\varepsilon^* U_\varepsilon(t, t_0)N_\varepsilon$ is block-diagonal with respect to the ε-independent decomposition given by P_0: $[P_0, N_\varepsilon^* U_\varepsilon(t, t_0)N_\varepsilon] = 0$ and the *reduced evolution* in $P_0\mathcal{H}$,

$$u_{\text{eff},\varepsilon} \equiv P_0 N_\varepsilon^* U_\varepsilon(t, t_0)N_\varepsilon P_0, \quad (7.4)$$

satisfies the equation

$$i\frac{d}{dt}u_{\text{eff},\varepsilon} = h_{\text{eff},\varepsilon}u_{\text{eff},\varepsilon}; \quad h_{\text{eff},\varepsilon} = P_0 N_\varepsilon^* H_\varepsilon N_\varepsilon P_0. \quad (7.5)$$

Also, the spectrum of H_ε enclosed by Γ coincides with the spectrum of $h_{\text{eff},\varepsilon}$ (as an operator in $P_0\mathcal{H}$). The point of the reduction process [Ka] outlined above, is that it allows us to replace part of the evolution problem (as well as spectral analysis of H_ε) in $\mathcal{H}$ with a reduced one (which is easier and more suited to analytical and numerical studies; e.g., in the one dimensional case the integration of (7.5) is trivial) in the smaller Hilbert space (in many instances finite-dimensional) $P_0\mathcal{H}$. Clearly, the crucial condition for such a reduction theory is the existence of an invariant subspace.

Unfortunately, the Rellich–Kato perturbation theory covers only a small part of the "perturbation" like problems of quantum mechanics, so an extension of the above reduction theory was needed. Gradually it become clear that the adiabatic theorem of quantum mechanics, the theory of spectral concentration, the theory of adiabatic invariants for linear Hamiltonian systems, the theory of simplifications and diagonalization of differential evolution equations and more recently singular perturbations as well as the semi-classical limit in quantum mechanics can be viewed as particular instances of a far reaching extension of the Rellich–Kato reduction theory. The price to be paid for such a generality is that one has to replace *invariant subspaces* by *almost invariant subspaces* and accordingly one of the central issues is to estimate the error.

The aim of this article is to present the basic facts, as well as some applications of this extension. Due to the space limitations we cannot cover, even partially, the very ramified topic of "singular perturbations" (adiabatic theorem and semi-classical limit included) in quantum mechanics and we apologize for not mentioning some other related important developments (for example, nothing will be said about Berry's phase, Landau–Zener formulae and magnetic perturbation theory). Also due to the very extended bibliography we would need, we shall give, whenever possible, mainly review papers from which additional bibliography can be traced.

The content of this article is as follows. In Section 7.2, we present the general setting and show that the existence of almost invariant subspaces leads, up to some controlled errors, to the same kind of results about evolutions as the existence of invariant subspaces. Also, the construction of almost invariant subspaces (more precisely the orthogonal projection onto those subspaces) out of some almost idempotents, $T_\varepsilon(t)$, is written down. The core of the theory is exemplified in Section 7.3 and consists in the "perturbative" construction of $T_\varepsilon(t)$ in various cases. The difficulty stems from the fact that the expansion in (7.2) can be at most asymptotic and the formula (7.2) does not hold. Accordingly, one has to find in each particular case an alternative formula for E_j.

7.2 Almost invariant subspaces: Generalities

A family, $\mathcal{L}_\varepsilon(t)$, of (closed) subspaces given by a family of orthogonal projections, $\mathcal{L}_\varepsilon(t) = P_\varepsilon(t)\mathcal{H}; \quad P_\varepsilon(t) = P_\varepsilon(t)^* = P_\varepsilon(t)^2$ is said to be *almost invariant* under the evolution $U_\varepsilon(t,t_0)$ given by (7.1) if

$$\| (1 - P_\varepsilon(t))U_\varepsilon(t,t_0)P_\varepsilon(t_0) \| \leq \omega(\varepsilon;t,t_0) \to_{\varepsilon\to 0} 0 . \tag{7.6}$$

In each particular case one has to find the time scales for which (7.6) holds true; the most usual are $|t - t_0| \sim \varepsilon^{-p}; \ e^{-c/\varepsilon}$. At the heuristic level the differential form of (7.6) is

$$\| i\frac{d}{dt}P_\varepsilon(t) - [H_\varepsilon(t), P_\varepsilon(t)] \| \leq \delta(\varepsilon;t) \to_{\varepsilon\to 0} 0 . \tag{7.7}$$

We take (7.7) as the *definition* of almost invariant subspaces and indeed one can show (see below) that (7.6) implies (7.7) with $\omega(\varepsilon;t,t_0) \leq \int_{t_0}^t \delta(\varepsilon;u)du$. Some remarks are in order here.

i. We would like to stress once again that $P_\varepsilon(t)$ have to be constructed from $H_\varepsilon(t)$ without any reference to $U_\varepsilon(t,t_0)$.

ii. Our presentation here is at a heuristic level but at the technical level, since in general $H_\varepsilon(t)$ are unbounded self-adjoint operators, one has to make precise the domain questions, existence of $U_\varepsilon(t,t_0)$, etc; in particular, in order (7.7) to make sense one needs a dense subspace $\mathcal{D}_0 \subset \mathcal{D}(H_\varepsilon(t))$ such that $P_\varepsilon(t)\mathcal{D}_0 \subset \mathcal{D}(H_\varepsilon(t))$. These matters have to be settled for each particular case at hand.

iii. Although true in some cases, in general it is not required that $H_\varepsilon(t)$ and/or $P_\varepsilon(t)$ have limits as $\varepsilon \to 0$.

The next object we introduce is the *approximate evolution* $U_{A,\varepsilon}(t,t_0)$ (called sometimes *superadiabatic evolution*), which is the generalization [Ne0, Ne4, Ne5, JP1] to the present setting of Kato's adiabatic evolution. More precisely $U_{A,\varepsilon}(t,t_0)$ is given as the solution of

$$i\frac{d}{dt}U_{A,\varepsilon}(t,t_0) = (H_\varepsilon(t) - B_\varepsilon(t))U_{A,\varepsilon}(t,t_0); \quad U_{A,\varepsilon}(t_0,t_0) = 1,$$

$$B_\varepsilon(t) = -(1 - 2P_\varepsilon(t))(i\frac{d}{dt}P_\varepsilon(t) - [H_\varepsilon(t), P_\varepsilon(t)]). \tag{7.8}$$

The point is that on one hand $P_\varepsilon(t)$ is invariant under $U_{A,\varepsilon}(t,t_0)$, i.e.,

$$P_\varepsilon(t) = U_{A,\varepsilon}(t,t_0)P_\varepsilon(t_0)U_{A,\varepsilon}(t,t_0)^* \tag{7.9}$$

and on the other hand $U_{A,\varepsilon}(t,t_0)$ is close to $U_\varepsilon(t,t_0)$:

$$U_\varepsilon(t,t_0) = U_{A,\varepsilon}(t,t_0)\Omega_{A,\varepsilon}(t,t_0); \quad \| \Omega_{\varepsilon,A}(t,t_0) - 1 \| \le \int_{t_0}^{t} \delta(\varepsilon,u)du. \tag{7.10}$$

Notice that (7.9), (7.10) imply (7.6) with $\omega(\varepsilon;t,t_0) \le \int_{t_0}^{t} \delta(\varepsilon,u)du$. The equality (7.9) means that the time-dependent family $P_\varepsilon(t)$ is *invariant* under $U_{A,\varepsilon}(t,t_0)$. In most cases it is possible to do better and to define a related evolution which leaves invariant a *fixed* subspace called *reference subspace*. More precisely *suppose* that one can find $W_\varepsilon(t)^* = W_\varepsilon(t)^{-1}$; $P_0 = P_0^* = P_0^2$ such that:

$$P_\varepsilon(t) = W_\varepsilon(t)P_0W_\varepsilon(t)^* .$$

Then one can write the analog of (7.5). Indeed, set (the "interaction" picture)

$$\tilde{U}_{\varepsilon,A}(t,t_0) = W_\varepsilon(t)^*U_{\varepsilon,A}(t,t_0)W_\varepsilon(t_0)$$

and verify by a simple computation that $[\tilde{U}_{\varepsilon,A}(t,t_0), P_0] = 0$, i.e., $\tilde{U}_{\varepsilon,A}(t,t_0)$ is block diagonal with respect to the decomposition given by P_0. Further, if one defines $u_{\text{eff},\varepsilon}(t,t_0) \equiv P_0\tilde{U}_{\varepsilon,A}(t,t_0)P_0$, then

$$i\frac{d}{dt}u_{\text{eff},\varepsilon}(t,t_0) = h_{\text{eff},\varepsilon}(t)u_{\text{eff},\varepsilon}(t,t_0), \quad u_{\text{eff},\varepsilon}(t_0,t_0) = 1 \tag{7.11}$$

with

$$h_{\text{eff},\varepsilon}(t) = P_0W_\varepsilon(t)^*(H_\varepsilon(t) - B_\varepsilon(t) - i(\frac{d}{dt}W_\varepsilon(t))W_\varepsilon(t)^*)W_\varepsilon(t)P_0 . \tag{7.12}$$

$h_{\text{eff},\varepsilon}(t)$ and $u_{\text{eff},\varepsilon}(t,t_0)$ are called *effective hamiltonian* and *effective evolution* respectively in the *reference subspace* $P_0\mathcal{H}$. Putting the things together one can write $U_\varepsilon(t,t_0)P_\varepsilon(t_0)$ in terms of $u_{\text{eff},\varepsilon}(t,t_0)$ up to errors less than $\int_{t_0}^{t} \delta(\varepsilon,u)du$:

$$U_\varepsilon(t,t_0)P_\varepsilon(t_0) = W_\varepsilon(t)u_{\text{eff},\varepsilon}(t,t_0)P_0 + r(t,t_0) \tag{7.13}$$

with

$$\| \, r(t, t_0) \, \| \leq \int_{t_0}^{t} \delta(\varepsilon, u) \, du \,. \tag{7.14}$$

With this our generalized reduction theory is complete. Of course all that is nothing but soft "general nonsense" since we are left with the construction of $P_\varepsilon(t)$ and $W_\varepsilon(t)$. While in most case one can find P_0 such that for sufficiently small ε, $\sup_{t \in \mathbb{R}} \| \, P_\varepsilon(t) - P_0 \, \| < 1$ and then $W_\varepsilon(t)$ can be taken to be the Sz-Nagy matrix intertwining $P_\varepsilon(t)$ and P_0, the construction of $P_\varepsilon(t)$ is, as already said, the hard core of the whole theory and we shall deal with this problem in the next section. Actually what will be constructed is not a family of orthogonal projections but a family of almost (self-adjoint) idempotents $T_\varepsilon(t)$ satisfying

$$\| \, T_\varepsilon(t)^2 - T_\varepsilon(t) \, \| \leq \eta(\varepsilon, t) < 3/16, \tag{7.15}$$

$$\| \, i\frac{d}{dt}T_\varepsilon(t) - [H_\varepsilon(t), T_\varepsilon(t)] \, \| \leq \tilde{\delta}(\varepsilon; t) \,. \tag{7.16}$$

Then the needed $P_\varepsilon(t)$ is given by the following Lemma [Ne4], [Ne5].

Lemma 1. *Suppose that $T_\varepsilon(t)$ is self-adjoint and satisfy (7.15) and (7.16). Then $P_\varepsilon(t)$ given by*

$$P_\varepsilon(t) \equiv \frac{i}{2\pi} \oint_{|z-1|=1/2} (T_\varepsilon(t) - z)^{-1} dz$$
$$= T_\varepsilon(t) + (T_\varepsilon(t) - 1/2)[(1 + 4(T_\varepsilon(t)^2 - T_\varepsilon(t)))^{-1/2} - 1] \tag{7.17}$$

satisfies

$$\| \, T_\varepsilon(t) - P_\varepsilon(t) \, \| \leq 40\eta(\varepsilon, t) \tag{7.18}$$

and

$$\| \, i\frac{d}{dt}P_\varepsilon(t) - [H_\varepsilon(t), P_\varepsilon(t)] \, \| \leq 8\tilde{\delta}(\varepsilon; t). \tag{7.19}$$

We end this section with a remark concerning the time-independent case, i.e., $H_\varepsilon(t) = H_\varepsilon$. In this case $P_\varepsilon(t) = P_\varepsilon$, $W_\varepsilon(t) = W_\varepsilon$ and (7.7) becomes

$$\| \, [H_\varepsilon, P_\varepsilon] \, \| \leq \delta(\varepsilon) \rightarrow_{\varepsilon \to 0} 0 \,. \tag{7.20}$$

Actually, by Howland's trick (known by physicists as the Floquet theory) one can consider only the time-independent case (at least as far as the construction of $P_\varepsilon(t)$) is considered). Indeed defining the hamiltonian

$$\mathbb{H}_\varepsilon = -i\frac{d}{dt} + H_\varepsilon(t), \tag{7.21}$$

and the almost invariant subspace

$$\mathbb{P}_\varepsilon = \int_{\mathbb{R}}^{\oplus} P_\varepsilon(t)\, dt \tag{7.22}$$

in the Hilbert space $\mathbb{K} = \int_{\mathbb{R}}^{\oplus} \mathcal{H}\, dt$, one observes that $-i\frac{d}{dt}P_\varepsilon(t) + [H_\varepsilon(t), P_\varepsilon(t)]$ can be rewritten as $[\mathbb{H}_\varepsilon, \mathbb{P}_\varepsilon]$.

7.3 Examples

In this section we give examples of "perturbative" construction of T_ε. The heuristic is as follows. We seek $P_\varepsilon(t)$ of the form

$$P_\varepsilon(t) \sim \sum_{j=0}^{\infty} \varepsilon^j\, E_j(t) \tag{7.23}$$

and try to determine $E_j(t)$ from the condition that the formal series in the r.h.s. of (7.23) solves (at the formal series level)

$$i\frac{d}{dt}P_\varepsilon(t) - [H_\varepsilon(t), P_\varepsilon(t)] \sim 0, \quad P_\varepsilon(t) \sim P_\varepsilon(t)^* \sim P_\varepsilon(t)^2\,. \tag{7.24}$$

As already said, since we are dealing with singular perturbations the series in the r. h. s. of (7.23) is not convergent (or even worse, only a finite number of E_j exist) so one has to truncate the series and define

$$T_\varepsilon(t) = \sum_{j=0}^{N_\varepsilon} \varepsilon^j\, E_j(t) \tag{7.25}$$

where N_ε has to be chosen for each problem at hand. From the physical point of view the most useful is to choose $N_\varepsilon = N$ *independent* of ε. This choice gives for the error $\delta(\varepsilon, t)$ the form

$$\delta(\varepsilon, t) = C_N(t)\, \varepsilon^{N+1}\,. \tag{7.26}$$

In view of physical applications, it is an important feature of the theory that the constants C_N can be computed explicitly (at least for low values of N). An example of such an estimation is given in [ANe] (see also [Te] for a discussion of this point). In many cases one can make a better choice, namely to take an ε-dependent truncation according to an optimal reminder estimate rule and this leads (under appropriate technical conditions on H_ε) to exponentially small errors

$$\delta(\varepsilon, t) = Ce^{-c/\varepsilon^\alpha}, \ C < \infty, \ c > 0, \ \alpha > 0. \tag{7.27}$$

However in this case the constants are much more difficult (if not impossible) to control.

7.3.1 Asymptotic perturbation theory for quantum mechanics

We send the interested reader to [Ne6, Ne2] for a detailed discussion and additional bibliography and to [Ne0, Ne3, Ne5] for proofs. Here we only recall that the results cover practically all singular perturbations interesting from the physical point of view (e.g., Stark and Zeeman effects in atoms and molecules and the Stark–Wannier ladder problem in solids) and that the perturbed subspaces are almost invariant up to exponentially long times, which in the present setting is the analog of the fact that the imaginary part of the corresponding resonances (in the cases when they can be defined) are exponentially small. Let us mention also that the idea of almost invariant subspaces has been used recently in a nonlinear context [GMS].

7.3.2 Adiabatic expansions

Most of the theory of almost invariant subspaces has been developed in connection with this subject and the reader is sent to [Ne1]–[Ne5], [MN2, ASY, AE, JP2] for reviews, further references and related results. Here we only mention that adiabatic expansion has been recently used [AES, ES] in the beautiful rigorous proof of the Kubo formula in the context of QHE and the fact that by Howland's trick mentioned above, the adiabatic expansions can be considered as particular cases of semi-classical expansions (see [Sj, MaNe] as well as the extended discussion in [Te]).

7.3.3 Time Dependent Perturbations

In this case [MN1]

$$H_\varepsilon(t) = H_0 + \varepsilon\, V(t) \tag{7.28}$$

where for simplicity we assume that $V(t)$ is uniformly bounded, $\parallel V(t) \parallel \leq v < \infty$. The most common examples of (7.28) are atoms or molecules subjected to external electromagnetic fields or spin systems in time variable magnetic fields. The "atomic" time-independent hamiltonian, H_0, is supposed to have a group, σ_0, of discrete almost degenerate eigenvalues, well separated from the rest of the spectrum. For a long time the basis for a perturbative computation of (7.1) was (Volterra–)Dyson series in the interaction picture. While the Dyson series is convergent for all times, the first nontrivial term gives a good approximation only for $|t - t_0| \lesssim 1/\varepsilon\, v$. On the other hand, for the particular case, $V(t) = \cos(\omega t) V$, if the distance between σ_0 and the rest of the spectrum is of order $\Delta \gg \omega$, it is known that resonant transitions between subspaces corresponding to σ_0 and $\sigma(H_0) \setminus \sigma_0$, respectively, become important only on time scales $|t - t_0| \sim 1/\lambda^{[\Delta/\omega]}$ and this suggests the existence of an almost invariant subspace up to this time scale. This phenomenon of breakdown of the usual perturbation theory on time scales much shorter than the time scale of the onset of resonant transitions was known as the appearance of "secular divergences" in quantum mechanics (for a detailed discussion as well as for a solution of the secular divergences problem in the particular case when σ_0 consists in a single nondegenerate eigenvalue see [LEK]). The existence of the almost invariant subspace below provides the solution to the problem of secular divergences in the general case i.e., when σ_0 consists of a group of discrete almost degenerate eigenvalues of total multiplicity N. In this case the effective hamiltonian is called the "dressed" N-level hamiltonian system. So much for the physical motivation and let us describe more precisely the result. Let $\sigma(H_0) = \sigma_0 \cup \sigma_1$,

$$V(t) = \int_\Omega d\omega \hat{V}(\omega) e^{-i\omega t}, \quad \int d\omega (1 + |\omega|) \parallel \hat{V}(\omega) \parallel < \infty,$$

$$\sigma_j(u) = \{E + u | E \in \sigma_j\}; \quad \omega^n = (\omega_1, \dots, \omega_{(n)}) \in \Omega^n \tag{7.29}$$

$$\Sigma_j(\omega^{(n)}) = \bigcup_{m=1}^n \sigma_j(\sum_{k=m}^j \omega_n) \cup \sigma_j; \quad d_n \equiv \inf_{\Omega^n} \mathrm{dist}(\Sigma_0(\omega^{(n)}), \Sigma_1(\omega^{(n)})).$$

While the above (complicated!) definition of d_n is the one needed in the proof of the result below, there is an alternative expression which is much more transparent from the physical point of view:

$$d_n = \inf\{|E_i - E_f - \sum_{j=1}^m \omega_j|, |E_i - E_f|\} \tag{7.30}$$

where the infimum is taken over $E_i \in \sigma_0$, $E_f \in \sigma_1$, $\omega_j \in \Omega$, $m = 1, 2, \dots, n$. From (7.30) one sees that d_n decreases with n. Moreover (7.30) makes clear the physical meaning of N_0 defined as the smallest integer for which $d_{N_0+1} = 0$: one needs at least $N_0 + 1$ "photons" for a resonant transition between σ_0 and σ_1. The explicit formula for $E_j(t)$ is contained in:

Lemma 2. *Let N_0 be the smallest integer for which $d_{N_0+1} = 0$ and for $\omega^{(j)} \in \Omega^j$, $j \leq N_0$, let $\Gamma(\omega^{(j)})$ be a contour enclosing $\Sigma_0(\omega^{(j)})$ but no points of $\Sigma_1(\omega^{(j)})$. Then for $1 \leq j \leq N_0$,*

$$E_j(t) = \frac{i}{2\pi} \oint_{\Gamma(\omega^{(j)})} dz \left\{ \prod_{m=1}^j (R_0(z + \sum_{p=m}^j \omega_p)\hat{V}(\omega_m)) \right\} R_0(z) \tag{7.31}$$

satisfy

$$E_j(t) = \sum_{k=0}^j E_{j-k}(t) E_k(t); \quad i\frac{d}{dt} E_j(t) = [H_\varepsilon(t), E_{j-1}(t)]$$

with $E_0 = P_0$ (the spectral projection of H_0 corresponding to σ_0).

Working out the procedure outlined in Section 7.2 one obtains almost invariant subspaces on time scale $|t - t_0| \lesssim 1/\lambda^{N_0}$. The result is optimal in the sense that on larger time scales, resonant transition sets in giving, in general, nonzero transition rates. Notice that in this case $\lim_{\varepsilon \to 0} \parallel P_\varepsilon(t) - P_0 \parallel = 0$ and the procedure outlined in Section 7.2 leads to effective dynamics in the reference subspace $P_0 \mathcal{H}$. For estimations as well as for the explicit computation of a few terms of the expansion in λ of the effective hamiltonian, we send the reader to [MN1]

7.3.4 The Semi-classical Limit

We shall consider here only the time-independent case. We take H_ε to be the Weyl quantization of a matrix or even operator-valued symbol on the phase space $\mathbb{R}^{2n}$, $\hat{h}(q, p; \varepsilon)$:

$$(H_\varepsilon \phi)(q) = \left(\mathrm{Op}_\varepsilon^w(\hat{h})\phi \right)(q)$$

$$= \frac{1}{(2\pi\,\varepsilon)^n} \int e^{i(q-u)p/\varepsilon} \hat{h}(\frac{q+u}{2}, p; \varepsilon)\phi(u)\, du\, dp. \tag{7.32}$$

We seek almost invariant subspaces corresponding to isolated parts of the spectrum of $\hat{h}(q, p; 0)$. We shall first outline [NS], at the heuristic level, the main points of the construction of almost invariant subspaces. We start by recalling (see, e.g., for details [Ma]) that if $A_\varepsilon = \mathrm{Op}_\varepsilon^w(\hat{a})$, $B_\varepsilon = \mathrm{Op}_\varepsilon^w(\hat{b})$, then $A_\varepsilon B_\varepsilon = \mathrm{Op}_\varepsilon^w(\hat{c})$ where $\hat{c}$ is given by the Weyl symbol product:

$$\hat{c}(q, p; \varepsilon) \equiv \widehat{(a \# b)}(q, p; \varepsilon) = \frac{1}{(2\pi\,\varepsilon)^{2n}} \int e^{i((p-\eta)(v-q)+(p-\theta)(q-v))/\varepsilon},$$

$$\hat{a}(\frac{q+u}{2}, \eta; \varepsilon)\hat{b}(\frac{q+v}{2}, \theta; \varepsilon) du\, dv\, d\eta\, d\theta\,.$$

Further if $\hat{a}$ is a semi-classical symbol, i.e., it has an asymptotic expansion $\hat{a}(q, p; \varepsilon) \sim \sum_{j=0}^{\infty} a_j(q, p)\,\varepsilon^j$, we say that $a(q, p; \varepsilon) \equiv \sum_{j=0}^{\infty} a_j(q, p)\,\varepsilon^j$ is the *formal* symbol of A_ε (or equivalently $\hat{a}(q, p; \varepsilon)$). Recall that the formal symbol does not define uniquely the symbol. The *formal Moyal–Weyl product* is defined by (here $D = i\partial$)

$$(a \# b)_j(q, p) = \sum_{|\alpha|+|\beta|+k+l=j} \frac{(-1)^{|\beta|}}{\alpha!\beta!2^{|\alpha|}2^{|\beta|}} \partial_p^\alpha D_q^\beta a_k(q, p)\partial_p^\beta D_q^\alpha b_l(q, p).$$

The formal symbol of the Weyl product equals the Moyal–Weyl product of the corresponding formal products. The vector space of formal symbols equipped with the $\#$ product is called the Moyal algebra. The very important feature of the formal product, $\#$, is that it is *local* in phase space, i.e., for each j, $(a \# b)_j$ depends only on a finite number of a_k, b_k and a finite number of their derivatives. The basic idea of finding almost invariant subspaces in semi-classical context is to construct first *Moyal projections* in the Moyal algebra, i.e., formal symbols

$$\pi(x, p; \varepsilon) \sim \sum_{j=0}^{\infty} \pi_j(q, p; \varepsilon) \, \varepsilon^j \tag{7.33}$$

satisfying (at a formal series level)

$$\pi \sim \pi^* \sim \pi \# \pi; \quad \pi \# h - h \# \pi \sim 0 \tag{7.34}$$

(here by $\pi^*(q, p; \varepsilon)$ we mean the adjoint of the matrix $\pi(q, p; \varepsilon)$). The lemma below says that one can always construct Moyal projections corresponding to isolated parts of the spectrum of the principal symbol $h_0(q, p)$ of $\hat{h}(q, p; \varepsilon)$: $\hat{h}(q, p; \varepsilon) = h_0(x, p) + \mathcal{O}(\varepsilon)$. Let $\sigma(q, p)$ be the spectrum of $h_0(x, p)$ and suppose that, for some $(q_0, p_0) \in \mathbb{R}^{2n}$, $\sigma(q_0, p_0) = \sigma_1(q_0, p_0) \cup \sigma_2(q_0, p_0)$,

$$\mathrm{dist}\,(\sigma_1(q_0, p_0), \sigma_2(q_0, p_0)) := d(q_0, p_0) > 0. \tag{7.35}$$

Then by perturbation theory there exists a neighborhood $\mathcal{U}(q_0, p_0)$ of (q_0, p_0) such that, for $(q, p) \in \mathcal{U}(q_0, p_0)$ the spectrum of $h_0(q, p)$ is well separated, i.e., (7.35) is satisfied on $\mathcal{U}(q_0, p_0)$ with $d(q, p) \geq d(q_0, p_0)/2$, and there exists a contour Γ_{q_0, p_0} (not depending upon $(q, p) \in \mathcal{U}(q_0, p_0)$) enclosing $\sigma_1(q, p)$ and satisfying

$$\mathrm{dist}\,(\Gamma_{q_0, p_0}, \sigma(q, p)) \geq d(q_0, p_0)/2 \,. \tag{7.36}$$

Lemma 3. *Suppose (7.35) holds true. Then, for $(q, p) \in \mathcal{U}(q_0, p_0)$, there exist unique $\pi_j(q, p)$, $j = 0, 1, \ldots$, such that $\pi_0(q, p)$ is the spectral projection of $h_0(q, p)$ corresponding to $\sigma_1(q, p)$ and the formal symbol $\sum_{j=0}^{\infty} \pi_j(q, p) \, \varepsilon^j$ satisfies (7.34).*

Lemma 3 appeared many times in the literature (see: [HS, EW, BN] and [Sj] where the earlier construction [Ne1, JP1] of adiabatic projections corresponding to the case $h(q, p; \varepsilon) = p + H_\varepsilon(q)$ was obtained in the framework of the theory of pseudo-differential operators). There are essentially two methods of proof. The first one [Ne1], [EW, BN], is a recurrent construction for π_j solving the equations coming from (7.34). The second one [HS],[Sj] is close in spirit to Rellich–Kato perturbation theory outlined in the introduction: due to (7.36), for all $z \in \Gamma_{q_0, p_0}$, one can construct (see, e.g., [Ma]) the parametrix

$$r(q, p; \varepsilon, z) \sim \sum_{j=0}^{\infty} r_j(q, p; z) \, \varepsilon^j$$

satisfying (at a formal series level)

$$r \# (h - z) \sim (h - z) \# r \sim 1 \tag{7.37}$$

and then obtain $\pi_j(q, p)$ from the Riesz formula for spectral projection

$$\pi_j(q,p) = \frac{i}{2\pi} \int_{\Gamma_{q_0,p_0}} r_j(q,p;z)\, dz. \tag{7.38}$$

In [NS] the second method has been used since, on the one hand, it gives explicit formulae for π_j and, on the other hand, allows a much easier control on $\pi_j(q,p)$ and their derivatives than the recurrent construction. Provided $\pi_j(q,p)$ can be globally defined (i.e., $\sigma_1(q_0,p_0)$ is isolated for all $(q,p) \in \mathbb{R}^n$) and are uniformly bounded together with their derivatives, by the Calderon–Vaillancourt theorem, one can define

$$E_j = \mathrm{Op}_\varepsilon^w(\pi_j). \tag{7.39}$$

Notice that in this case E_j themselves depend on ε. As an example we consider two-component Klein–Gordon systems [NS]. For further applications, extensions and extensive bibliography on older and more recent developments see [Te] (also the article by S. Teufel in this volume). The two-component Klein–Gordon hamiltonian is

$$H_\varepsilon = a(\varepsilon\, D_q)\mathbf{1}_2 + V(q) = \mathrm{Op}_\varepsilon^w(a(p)\mathbf{1}_2 + V(q)) \tag{7.40}$$

acting on $\mathcal{H} = L^2(\mathbb{R}^n)^{\oplus 2} = L^2(\mathbb{R}^n) \oplus L^2(\mathbb{R}^n)$. Here $a(p) = (\sqrt{p^2+1}-1)$, $\mathbf{1}_2$ is the 2×2 identity matrix and $V(q)$ is a 2×2 hermitian matrix-valued function that admits an analytic extension in some strip $I_a = \{q \in \mathbb{C}^n \ ; \ |\mathrm{Im}q| < a\}$, $a > 0$. Since we have in mind applications to scattering theory we assume that $V(q)$ to be "short range", i.e., there exists $V_\infty = \begin{pmatrix} \lambda_{1,\infty} & 0 \\ 0 & \lambda_{2,\infty} \end{pmatrix}$ such that

$$\sup_{q \in I_a} \langle q \rangle^\delta |V(q) - V_\infty| < \infty; \quad \langle q \rangle = \sqrt{1+q^2} \tag{7.41}$$

with $\delta > 1$. Concerning the spectrum of $V(q)$, we assume that the two eigenvalues $\lambda_1(q) > \lambda_2(q)$, satisfy

$$\inf_{q \in I_a} |\lambda_1(q) - \lambda_2(q)| = 2d > 0. \tag{7.42}$$

It follows that on I_a,

$$V(q) = \lambda_1(q)\pi_0(q) + \lambda_2(q)\,(1 - \pi_0(q))$$

where $\pi_0(q)$ is a bounded projection in $\mathbb{C}^2$, analytic in I_a and self-adjoint for $q \in \mathbb{R}^n$. Moreover, we suppose that $V_{ij}(q) = \overline{V_{ij}(q)}$. In what follows, for an operator A and $\eta > 0$ we shall denote

$$\| A \|_\eta = \sup_{t \in [0,1]} \| \langle x \rangle^{t\eta} A \langle x \rangle^{(1-t)\eta} \|.$$

Under the above conditions one can prove [NS] that H_ε has an almost invariant subspace and is unitarily equivalent with an almost diagonal effective hamiltonian.

Theorem 1. H_ε *is unitarily equivalent with*

$$H_{\text{eff},\varepsilon} = \begin{pmatrix} a(\epsilon D_q) + \lambda_1(q) & 0 \\ 0 & a(\epsilon D_q) + \lambda_2(q) \end{pmatrix} + \epsilon B_\varepsilon \qquad (7.43)$$

where $B_{\varepsilon,ij};\ \ i,j = 1,2$ are bounded operators satisfying

$$\|B_{\varepsilon,ii}\|_\delta = \mathcal{O}(1); \ \ \|B_{\varepsilon,12}\|_\delta = \mathcal{O}(e^{-c/\epsilon}) \qquad (7.44)$$

uniformly with respect to sufficiently small ϵ.

We turn now to applications to spectral and scattering theory for H_ε. Consider the scattering matrix, S, corresponding to the pair H_ε and

$$H_{\varepsilon,0} = a(\epsilon D_q)\mathbf{1}_2 + \begin{pmatrix} \lambda_{1,\infty} & 0 \\ 0 & \lambda_{2,\infty} \end{pmatrix}. \qquad (7.45)$$

Theorem 2. *Let, for $j = 1,2$:*

$$H_{\text{eff},\varepsilon,j} = a(\epsilon D_q) + \lambda_j(q) + \epsilon B_{\varepsilon,ii}; \quad H_{0,\varepsilon,j} = a(\epsilon D_q) + \lambda_{j,\infty} \qquad (7.46)$$

acting in $L^2(\mathbf{R}^n)$ and S_j the scattering operator corresponding to the pair $H_{\text{eff},\varepsilon,j}, H_{0,\varepsilon,j}$. Then, there exists $c > 0$-independent of ϵ, such that, for any (here $\hat\phi$ denotes the Fourier transform) $\phi_k \in \mathcal{H}$, $\hat\phi_k \in \mathcal{C}_0^\infty(\mathbf{R}^n\backslash\{0\})^{\oplus 2}, k = 1,2$ and $\epsilon \in (0, \epsilon_0]$ sufficiently small, we have

$$|(\phi_2, \left(S - \begin{pmatrix} S_1 & 0 \\ 0 & S_2 \end{pmatrix}\right)\phi_1)| \le C(\phi_1,\phi_2)c^{-c/\epsilon} \qquad (7.47)$$

for some constant $C(\phi_1, \Phi_2) > 0$ depending on ϕ_1, ϕ_2 but not on ϵ.

Since S commutes with $H_{\varepsilon,0}$, one can define the scattering matrix at fixed energy $S(E)$. Let us stress that Theorem 2 does not imply the corresponding result for $S(E)$; when shrinking the supports of ϕ_1 and ϕ_2 the constant $C(\phi_1,\phi_2)$ may blow up. At the heuristic level the reason for this blow up is clear: Theorem 1 (via the Duhamel formula) implies that the transitions *per unit time* between the two levels is exponentially small, but if, at the given energy, there is a resonance very (exponentially) close to the real axis, it has an exponentially long life time and during this time considerable transitions can take place. Accordingly, one expects exponentially block diagonalization of $S(E_0)$ only if there are no resonances around E_0 or in other words if $S(E)$ is smooth in a neighborhood of E_0.

We turn now to some direct applications of Theorem 1 to spectral theory. We shall discuss $H_{\text{eff},\varepsilon}$ but all the results apply to H_ε also, since they are unitarily equivalent. If $\inf_{x\in\mathbb{R}^n} \lambda_2(x) := m_2 < \lambda_{2,\infty}$, then $H_{\text{eff},\varepsilon}$ has discrete spectrum in the interval $(m_2, \lambda_{2,\infty})$ which, up to exponentially small errors, equals the union of the (discrete) spectra of H_ε in this interval (see [Ne5] for details). Suppose now $\inf_{x\in\mathbb{R}^n} \lambda_1(x) := m_1 < \lambda_{1,\infty}$. Then $H_{\text{eff},\varepsilon,1}$ has bound states in the interval $(m_1, \lambda_{1,\infty})$ and as a consequence the diagonal part of $H_{\mathrm{eff},\varepsilon}$ has bounded states embedded in the continuum spectrum in the interval $(\max\{m_1, \lambda_{2,\infty}\}, \lambda_{1,\infty})$. The small off-diagonal term turns (generically) these bound (stable) states into metastable states. Let Ψ be an eigenfunction of such a bound state, i.e.,

$$H_{\mathrm{eff},\varepsilon,1}\Psi = E\Psi, \quad E \in (\max\{m_1, \lambda_{2,\infty}\}, \lambda_{1,\infty}).$$

Then from Theorem 1

$$\|(H_{\mathrm{eff},\varepsilon} - E)\Psi\| \leq Ce^{-c/\epsilon},$$

i.e., Ψ, E are pseudo-eigenvectors and pseudo-eigenvalues (see [Ka]) of exponential order for $H_{\mathrm{eff},\varepsilon}$ (the name quasi-modes is also used). A more physical picture is given by Duhamel's formula which together with Theorem 1 gives

Corollary 3

$$|(\Psi, e^{-itH_{\mathrm{eff},\varepsilon}}\Psi)| \geq 1 - C|t|e^{-c/\epsilon}. \tag{7.48}$$

As it stands Corollary 3 gives the *existence* and the control on the (exponentially long) life time of metastable states (quasi-modes) in the semi-classical limit. However in many instances one can make the connection with the resonances (defined as poles of the resolvent). In the cases when this connection can be made, Corollary 3 says that the imaginary part of the resonances is exponentially small.

We end this subsection by once again calling the reader's attention to the recent review [Te] where the semi-classical limit together with applications not covered here is extensively discussed. In particular applications to the dynamics of Bloch and Dirac electrons subjected to slowly varying external potentials are considered. Our remark here is that, while in order to deal with nice classes of symbols one has to assume that the external potentials have to be bounded, a closer look at the particular structure of the symbols involved allows us to treat (without any additional technicalities) the case when only the fields (i.e., the derivatives of the potentials) have to be uniformly bounded (a condition with a clear physical meaning).

References

[AES]   M. Aizenman, A. Elgart, G.H. Schenker, Adiabatic charge transport and the Kubo formula for 2D Hall conductance, preprint.

[ANe]   A. Nenciu, Spectrul electronilor Bloch in prezenta campurilor externe slab neuniforme, *Studii si Cercetari de Fizica* **39** (1987), 494–545.

[ASY]   J. Avron, R. Seiler, L.G. Yaffe, Adiabatic theorems and applications to the QHE, *Commun. Math. Phys.* **110** (1987), 33–49.

[AE]   J. Avron, A. Elgart, Adiabatic theorem without a gap condition, *Commun.Math. Phys.* **203** (1999), 445–463.

[BN]   R. Brummelhuis, J. Nourrigat, Scattering amplitude for Dirac operators, *Commun. Partial Diff. Equations* 24 (1999), 377–394.

[ES]   A. Elgart, G.H. Schenker, Adiabatic charge transport and the Kubo formula for Landau type hamiltonians, preprint.

[EW]   C. Emmerich, A. Weinstein, Geometry of transport equation in multicomponent WEB approximations, *Commun. Math. Phys.* **176** (1996), 701–711.

[GMS]   V.Grecchi, A.Martinez, A.Sacchetti, Destruction of the beating effect for a non-linear Schrödinger equation, *Commun. Math. Phys.* **227** (2002), 191–209.

[HS]   B. Helffer, J. Sjöstrand, Analyse semiclassique pour l'équation de Harper II, *Mem. Soc. Math. France., Nouv. Serie.* **40** (1990), 1–139.

[JP1]   A. Joye, Ch-E. Pfister, Superadiabatic evolution and adiabatic transition probability between two non-degenerate levels isolated in the spectrum, *J. Math. Phys.* **34 (1993)**, 454–479.

[JP2]   A. Joye, Ch-E. Pfister, Exponential estimates in adiabatic quantum evolutions. In: D.E.Witt, A.J.Bracken, M.D.Gould, P.A.Pearce, eds., *XIIth International Congress of Mathematical Physics (ICMP'97)*.

[Ka]   T. Kato, *Perturbation Theory for Linear Operators*, 2nd ed., Classics in Mathematics, Springer-Verlag, Berlin, 1980.

[LEK]   P.W. Langhoff, S.T. Epstein, M. Karplus, *Rev. Mod. Phys.* **44** (1972), 602.

[MN1]   Ph-A. Martin, G. Nenciu, Perturbation theory for time dependent hamiltonians: rigorous reduction theory, *Helv. Phys. Acta* **65** (1992), 528–559.

[MN2]   Ph-A. Martin, G. Nenciu, Semi-classical inelastic S-matrix for one-dimensional N-states systems, *Rev. Math. Phys.* **7** (1995), 193–242.

[Ma]   A. Martinez, *An Introduction to Semiclassical and Microlocal Analysis*, Springer, Berlin, 2001.

[MaNe]   A. Martinez, G. Nenciu, On adiabatic reduction theory, *Oper. Theory: Ad. Appl.* **78** (1995), 243–252.

[Ne0]   G. Nenciu, Adiabatic theorem and spectral concentration, *Commun. Math. Phys.* **82** (1981), 121–135.

[Ne1]   G. Nenciu, Asymptotic invariant subspaces, adiabatic theorems and block diagonalization, in *Recent Developments in Quantum Mechanics*, Boutet de Monvel et al., eds., Kluwer, Dordrecht, 1991.

[Ne2]   G. Nenciu, Dynamics of band electrons in electric and magnetic fields: rigorous justification of the effective hamiltonians, *Rev. Mod. Phys.* **63** (1991), 91–128.

[Ne3]   G. Nenciu, Linear adiabatic theory; exponential estimates, *Commun. Math. Phys.* **152** (1993), 479–496.

[Ne4]   G. Nenciu, Linear adiabatic theory: Exponential estimates and applications, in *Algebraic and Geometric Methods in Mathematical Physics* Boutet de Monvel and V. Marcenco, eds., Kluwer, Dordrecht, 1996.

[Ne5]   G. Nenciu, On asymptotic perturbation theory for quantum mechanics: almost invariant subspaces and gauge invariant magnetic perturbation theory, *J. Math. Phys.* **43** (2002), 1273–1298.

[Ne6]   G. Nenciu, On asymptotic perturbation theory for quantum mechanics, in *Long Time Behavior of Classical and Quantum Systems*, S. Graffi and A. Martinez, eds.,World Scientific, Singapore, 2001.

[NS]   G. Nenciu, Vania Sordoni, Semi-classical limit for multistage Klein-Gordon systems: almost invariant subspaces and scattering theory. Preprint mp-arc 01-36 (2001).

[Sj]   J. Sjöstrand, *Projecteurs adiabatiques du point de vue pseudodifférentiel*, *C. R. Acad. Sci. Paris* **317**, Série I (1993), 217–220.

[Te]   S. Teufel, *Adiabatic Perturbation Theory in Quantum Dynamics*, Habilitationsschrift, Zentrum Mathematik, Technische Universität München, 2002.

Nonlinear Asymptotics for Quantum Out-of-Equilibrium 1D Systems: Reduced Models and Algorithms

F. Nier and M. Patel

IRMAR, UMR-CNRS 6625
Campus de Beaulieu
Université de Rennes I
F-35042 Rennes Cedex

8.1 Introduction

8.1.1 The problem

Although the mathematical analysis can be extended to other models, our main motivation is the understanding of the nonlinear dynamics of 1D quantum electronic devices like resonant tunneling diodes and superlattices. The final aim is the development of reduced models and algorithms in order to compute and predict the behavior of such devices. Experiments (reported a.e. in [3]) show rather complex nonlinear phenomena, with hysteresis and steadily oscillating currents, which have been only partially elucidated up to now.

We are interested in the nonlinear dynamics of a beam of particles experimenting a localized nonlinear interaction in a potential structure like the one in Fig. 8.1. Before writing the complete system, here is the mathematical program that can be followed for such a nonlinear problem:

1) Functional framework: existence, uniqueness and stability under perturbation (done in [1][2][12]).
2) Determination of steady states (studied in [8] [13] and present work).
3) Large time asymptotics, nonlinear stability and instability of steady states, attractors (studied in [8] [13]).
4) Control theory (unstable steady states might be the most interesting from a technological point of view).

The initial problem is modelled by an infinite-dimensional dynamical system and in order to develop the above program as accurately as possible one can try to reduce it to some finite-dimensional one. Here we are led by the intuition of physicists who claim that the nonlinear dynamics of such systems is governed by a finite number of quantum resonant states. The model should,

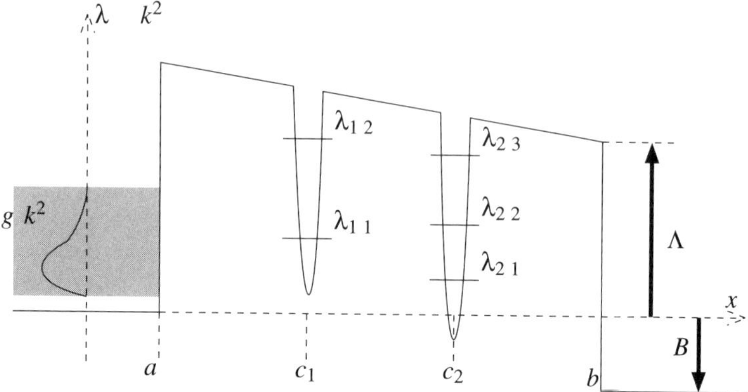

Fig. 8.1. Potential: Linear part $\mathcal{V}_0^h + \mathcal{B}$.

at least in some asymptotic meaning, follow this intuition and the more advanced work in this direction up to now is the one by Jona-Lasinio, Presilla and Sjöstrand in [8] and [13]. Here we will be slightly more specific about the asymptotics in order to develop a full mathematical asymptotic analysis of the nonlinear problem.

8.1.2 The model

Here are the characteristics of the model:

1) Electronic repulsion is described in a Hartree approximation valid for 1D systems and also called a Schrödinger–Poisson system.
2) Out of equilibrium steady states in 1D, can be described as functions of the asymptotic momentum $f(P_-) = \text{s-lim}_{t\to-\infty} e^{itH} f(\frac{x}{t}) e^{-itH}$ and involve scattering quantities whereas equilibrium states are function of the energy (see [12] for a more general presentation).
3) In order to have a finite of number of quantum resonant states asymptotically split from the rest of the continuous spectrum (that is very close to the real axis), one can consider the situation of quantum wells in a semiclassical island.

The small parameter $h > 0$ can be introduced after a rescaling under the assumption that the potential barriers are wide or high. The quantum hamiltonian for a single electron has the form

$$H^h(x, hD_x) = (hD_x)^2 + \mathcal{B}(x) + \mathcal{V}_0^h(x) + V_{NL}^h(x), \quad D_x = \frac{1}{i}\partial_x, \qquad (8.1)$$

with a nonlinear potential $V_{NL}^h(x)$ (which is nonnegative). The first potential term $\mathcal{B}(x)$ simply includes the bias voltage applied to the device. It is piecewise affine

$$\mathcal{B}(x) = -B \left[\frac{x-a}{b-a} 1_{[a,b]}(x) + 1_{[b,+\infty)}(x) \right].$$

The second term describes the barriers and the quantum wells,

$$V_0^h(x) = V_0 1_{[a,b]}(x) + \sum_{j=1}^{N} W_j \left(\frac{x-c_j}{h} \right),$$

with the constants $V_0 > 0$, $a < c_1 < c_2 < \cdots < c_N < b$ and the compactly supported potentials $W_j \in L^\infty(\mathbb{R})$, $W_j \leq 0$, fixed. When the potentials $-W_j$, $j \in \{1, \ldots, N\}$, are bounded from above by V_0, the hamiltonian H^h has only absolute continuous spectrum and a function of the asymptotic momentum can be written in terms of generalized eigenfunctions:

$$f(P_-^h)(x,y) = \int_{\mathbb{R}} f(k) \psi_-^h(k,x) \overline{\psi_-^h(k,y)} \, \frac{dk}{2\pi h}. \tag{8.2}$$

In our specific one-dimensional situation, the incoming generalized eigenfunctions $\psi_-(k,x)$ are given for $k > 0$ by

$$H^h(x, hD_x)\psi_-^h(k,x) = k^2 \psi_-(k,x) \quad \text{for } x \in \mathbb{R}, \tag{8.3}$$

$$\psi_-^h(k,x) = e^{ikx/h} + R(k)e^{-ikx/h} \quad \text{for } x \leq a \tag{8.4}$$

and for $k < 0$, $k^2 \neq B$, by

$$H^h(x, hD_x)\psi_-^h(k,x) = (k^2 - B)\psi_-(k,x) \quad \text{for } x \in \mathbb{R}, \tag{8.5}$$

$$\psi_-^h(k,x) = e^{ikx/h} + R(k)e^{-ikx/h} \quad \text{for } x \geq b. \tag{8.6}$$

The shape of the incoming beam of electrons is contained in the given function f. A beam coming from the left-hand side is described by a function f supported in $k \geq 0$ and we assume the form

$$f(k) = g(k^2)1_{\mathbb{R}_+}(k), \quad \text{with } 0 \leq g \leq g_{\max} < +\infty.$$

The density n^h of particle associated with $f(P_-)$ is then defined as a positive measure by duality:

$$\forall \varphi \in C_c^0((a,b)), \quad \int_a^b \varphi(x) \, dn^h(x) = \text{Tr} \left[g((P_-^h)^2)1_{\mathbb{R}_+^*}(P_-^h)\varphi(x) \right]. \tag{8.7}$$

It equals

$$n^h(x) = \int_0^{+\infty} g(k^2) \left| \psi_-^h(k,x) \right|^2 \frac{dk}{2\pi h}.$$

Finally the nonlinear potential V_{NL}^h is obtained as the solution of the Poisson equation

$$\begin{cases} -\partial_x^2 V_{NL}^h(x) = n^h(x), & \text{in } (a,b) \\ V_{NL}^h(a) = V_{NL}^h(b) = 0. \end{cases} \tag{8.8}$$

8.2 Mathematical asymptotic analysis

The existence of steady states for the system (8.1)(8.7)(8.8) for a fixed $h > 0$ was proved in [1] [2] and [12]. We consider the asymptotics $h \to 0$ in order to derive an easily solvable model involving only the real part of resonances which are in a finite number.

8.2.1 Assumptions, notation

The assumptions are:

Hypothesis 8.2.1
a) The number of quantum wells N is finite, $a < c_1 < \cdots < c_N < b$ and the potentials W_j, $j \in \{1, \ldots, N\}$ are bounded compactly supported with

$$0 \leq W_j \leq V_0, \quad V_0 > 0.$$

b) The quantity $\Lambda = V_0 - B$ is positive.
c) The function g is nonnegative and belongs to $C_c^0((0, \Lambda))$, the space of compactly supported continuous functions in $(0, \Lambda)$.
d) For some small constant $0 < c_g \preceq 1$, c_g determined by g, the distance between the wells satisfies $c_N - c_1 < c_g \min\{(c_1 - a), (b - c_N)\}$.[1]

For technical reasons, we assume in c) that the energy support of the incoming beam avoids the energy threshold 0. The situation where g is a decaying function on $(0, +\infty)$ is recovered afterwards by considering the limit $\varepsilon \to 0$ for a family of functions $g^\varepsilon(\lambda) = \chi(\varepsilon^{-1}\lambda)g(\lambda)$ with $\operatorname{supp} \chi \subset (0, +\infty)$, $\chi \equiv 1$ for $\lambda \geq 1$.

For $j \in \{1, \ldots, N\}$, we introduce the finite set $\{E_{j,k},\ 1 \leq k \leq K(j)\}$ of negative eigenvalues of the one-well quantum hamiltonian $-\partial_x^2 + W_j(x)$, labelled in the increasing order. (In dimension $d = 1$ these eigenvalues are simple.) The set of all these energy levels is $\mathcal{L} = \cup_{j=1}^{N} \{1, \ldots, K(j)\}$. For $l = (j, k) \subset \mathcal{L}$, we set $j(l) = j$, $k(l) = k$ and the quantity

$$\lambda_l = V_0 + \mathcal{B}(c_{j(l)}) + E_l \tag{8.9}$$

---

[1] Such an assumption ensures that the resonances of $H^h(x, hD_x)$ involved in the nonlinear problem are simple and can be treated individually even if there is some interaction between the wells according to the universal lower bound of the splitting for 1-dimensional hamiltonians given in [9]. Without this assumption one encounters the problem of possible multiple or accumulating resonances as in [5] with Jordan blocks or ill-conditioned diagonalization. Although this difficulty should be solved or bypassed in the present problem, we were not able to do it right now.

denotes the energy level shifted along the energy axis due to the classical linear potential $V_0 1_{[a,b]}(x) + \mathcal{B}(x)$. The density n^h will lie in the space $\mathcal{M}_b((a,b))$ of bounded measures on (a,b) and the nonlinear potential V_{NL}^h in

$$BV_0^2([a,b]) = \{V \in \mathcal{C}^0([a,b]), \ V'' \in \mathcal{M}_b((a,b)), \ V(a) = V(b) = 0\}.$$

This space is continuously embedded in the space of Lipschitz functions $W^{1,\infty}([a,b])$ and compactly embedded in the Hölder space $\mathcal{C}^{0,\alpha}([a,b])$ for any $\alpha < 1$. We will prove that the possible asymptotic potentials stand in the finite-dimensional subspace

$$\mathbb{P}_0^1[\mathbf{c}] = \left\{ V \in \mathcal{C}^0([a,b]), \ V(a) = V(b) = 0, V \Big|_{[c_j, c_{j+1}]} \text{ affine for } j \in \{0, \ldots, N\} \right\},$$

$$\text{with } \mathbf{c} = \{c_0 = a, c_1, \ldots, c_N, c_{N+1} = b\}.$$

For a nonnegative potential $V \in BV_0^2([a,b])$, $V \geq 0$, and an energy $\lambda \in (0, \Lambda)$, we introduce the Agmon distance associated with $\mathcal{V}^h(x) = V_0^h(x) + \mathcal{B}(x) + V(x)$,

$$\forall a \leq x \leq y \leq b, \quad d_{\mathrm{Ag}}(x, y; \lambda, V) = \int_x^y \sqrt{V_0^h(t) + \mathcal{B}(t) + V(t) - \lambda} \ dt.$$

$$(8.10)$$

The possible asymptotics of the potential V_{NL}^h will be discussed according to the quantities $\delta_-(l, V)$, $\delta_+(l, V)$ defined below for $l \in \mathcal{L}$, $V \in BV_0^2([a,b])$. We set first

$$j_M(l) = \max \{j(l'), \quad \lambda_{l'} + V(c_{j(l')}) = \lambda_l + V(c_{j(l)}), \ l' \in \mathcal{L}\} \quad (8.11)$$

$$\text{and } j_m(l) = \min \{j(l'), \quad \lambda_{l'} + V(c_{j(l')}) = \lambda_l + V(c_{j(l)}), \ l' \in \mathcal{L}\}. \quad (8.12)$$

Then $\delta_-(l, V)$ and $\delta_+(l, V)$ are respectively defined for $\lambda_l + V(c_j(l)) \in (0, \Lambda)$ by:

$$\delta_-(l, V) = d_{\mathrm{Ag}}(a, c_{j_m(l)}; *) - d_{\mathrm{Ag}}(c_{j_m(l)}, b; *) \quad (8.13)$$

$$\text{and } \delta_+(l, V) = d_{\mathrm{Ag}}(c_{j_M(l)}, b; *) - d_{\mathrm{Ag}}(a, c_{j_M(l)}, *), \quad (8.14)$$

where $*$ stands here for $(\lambda_l + V(c_{j(l)}), V)$. We have

$$\delta_+(l, V) \leq -\delta_-(l, V)$$

with equality only when $j_m(l) = j_M(l)$, which means

$$\{l' \in \mathcal{L}, \lambda_{l'} + V(c_{j(l')}) = \lambda_l + V c_{j(l)}\} = \{l\}.$$

We end this set of notation with

Definition 1. *The set of nonnegative potentials* $\mathbb{P}_0^1[\mathbf{c}]_+ = \{V \in \mathbb{P}_0^1[\mathbf{c}], \ V \geq 0\}$ *is partitioned into two subsets*

$$\mathcal{C} = \{V \in \mathbb{P}_0^1[\mathbf{c}]_+, \forall l \in \mathcal{L}, (0 < \lambda_l + V(c_{j(l)}) < \Lambda)$$
$$\Rightarrow (\delta_-(l, V) > 0 \text{ or } \delta_+(l, V) > 0)\},$$
$$\mathcal{Q} = \mathbb{P}_0^1[\mathbf{c}]_+ \setminus \mathcal{C}.$$

8.2.2 Main results

Here are the main mathematical results.

Theorem 8.2.2 *The solutions* V_{NL}^h, $h \in (0,1]$, *describe a bounded set of*

$$BV_0^2([a,b])_+ = \left\{ V \in BV^2([a,b]), V \geq 0, V(a) = V(b) = 0 \right\}$$

and thus is relatively compact in $\mathcal{C}^{0,\alpha}([a,b])$ *for any* $\alpha < 1$. *The set of* $\mathcal{A}$ *of its limit points as* $h \to 0$ *is included in* $\mathbb{P}_0^1([a,b])_+$. *More precisely, any* $V \in \mathcal{A}$ *solves*

$$-\partial_x^2 V = \sum_{l \in \mathcal{L}} t_l g\left(\lambda_l + V(c_{j(l)})\right) \delta_{c_{j(l)}}, \quad V(a) = V(b) = 0, \tag{8.15}$$

where the coefficients $t_l, l \in \mathcal{L}$, *satisfy*

$$0 \leq t_l \leq 1, \ (\delta_-(l,V) > 0) \Rightarrow (t_l = 1) \ and \ (\delta_+(l,V) < 0) \Rightarrow (t_l = 0). \tag{8.16}$$

The possible limits lying in $\mathcal{C}$ can be given by a variational formulation using

$$G(\lambda) = -\int_\lambda^{+\infty} g(s) \ ds.$$

Theorem 8.2.3 *The set* $\mathcal{A} \cap \mathcal{C}$ *is given by the collection of critical points in* $\mathbb{P}_0^1([a,b])_+$ *for the functionals*

$$J_{\mathcal{K}}(V) = \frac{1}{2} \int_a^b |\partial_x V(x)|^2 \ dx - \sum_{l \in \mathcal{K}} G(\lambda_l + V(c_{j(l)})), \quad \mathcal{K} \subset \mathcal{L}, \tag{8.17}$$

which satisfy the compatibility condition

$$\mathcal{K} = \left\{ l \in \mathcal{L}, \ 0 < \lambda_l + V(c_{j(l)}) < \Lambda \ and \ \delta_-(l,V) > 0 \right\}.$$

Included in some bootstrap process, the strategy of the proof is the following:

a) One first obtains uniform estimates for the density $n^h \in \mathcal{M}_b((a,b))$ with two monotonicity principles:

First monotony principle: Any bounded function of the asymptotic momentum $f(P_-^h)$ can be estimated by a decaying function of the energy

$$f(P_-^h) \leq \chi(H^h(x, hD_x)) \quad \text{if } f(k) \leq \chi(\mathcal{E}(k)),$$

with $\mathcal{E}(k) = k^2$ for $k \geq 0$ and $\mathcal{E}(k) = k^2 - B$ for $k < 0$. Hence the particle density n^h is estimated by the density (8.7) $\tilde{n}^h$ defined by

$$\forall \varphi \in \mathcal{C}_c^0((a,b)), \quad \int_a^b \varphi(x) \ d\tilde{n}^h(x) = \text{Tr} \left[\chi(H^h(x, hD_x))\varphi(x) \right].$$

Second monotonicity principle: If one calls $\tilde{n}_V$ the density defined above for a Hamiltonian $H_V = H_0 + V(x)$ for a decaying function χ, one has the convexity inequality (see a.e. [11] and [10])

$$\int_a^b \left[\tilde{n}_{V_2} - \tilde{n}_{V1}\right] \left[V_2 - V_1\right] \, dx \leq 0.$$

b) Linear analysis: Adapt Helffer–Sjöstrand's theory of resonances (see [7]) to the case of a Lipschitz potential in 1D with a systematic use of Grushin's problem (generalized and more flexible version of the Feshbach method) and Agmon distance (action) in order to describe accurately the effect of resonances close to the real axis on the quantities $\tilde{n}^h$ and n^h. A polar expansion of the local density of states which can be adapted from [4] leads to an accurate approximation of $\tilde{n}^h$. The derivation of a similar asymptotic formula for n^h requires a similar development for functions of the asymptotic momentum as for function of the energy. The discussion about the comparison between Agmon distance to $x = a$ and to $x = b$ enters at this last point.

8.3 Applications

8.3.1 Comments

One may question the validity of this asymptotic modelling compared to real situations. Indeed in actual electronic devices the widths of the wells and of the barrier have the same order of magnitude. Here are the answer that we give to justify such an analysis.

a) Indeed the introduction of the small parameter is not motivated by an exact geometrical scaling but should try to fit the important physically relevant quantities which are the positions of the real part of resonances. Here is the way to adjust the scaling to the physically realistic case: 1) Compute the resonances for the real potential with $B = 0$ and $V_{NL} = 0$. 2) Replace the real hamiltonian by an h-dependent hamiltonian (8.1) chosen so that the real part of resonances coincide for $B = 0$ and $V_{NL} = 0$ and so that the c_j's equal the middle position of the wells. 3) Solve the asymptotic nonlinear problem (8.15).
b) The previous method is not exact but implements the idea that the non-linear phenomena are governed by a finite number of resonances. The intuition of physicists is certainly correct, because the parameter h does not need to be very small (or in more physically realistic interpretation, the barriers do not need to be very wide or high) so that the resonances are very close to the real axis (remember that the imaginary part of resonances are of order $e^{-2S_0/h}$ for some Agmon distance S_0.).

c) As will be explained below for one or two wells, the number of nonlinear steady states rely on topological and geometrical arguments which should be robust under the modification proposed in a).

d) It is certainly hard for two reasons to find all the nonlinear steady states by numerical computations after discretisation of the full quantum problem: 1) The discretisation of the continuous spectrum may miss the narrow energy window of order $e^{-2S_0/h}$ were the quasiresonances lie. 2) As shown in the simple 2-wells case below, the number of nonlinear solutions may be quite high (up to 7) and any efficient nonlinear algorithm has to start close enough to a solution to find it. Hence if one wants accurate and realistic numerical results, the strategy could be the following: 1) Compute all the possible nonlinear solutions of the asymptotic model (8.15) according to the algorithm below; 2) use these solutions as initial guesses in a numerical simulation of the full quantum problem.

e) The variational formulation (8.17) suggests a way to analyze the nonlinear stability of $\mathcal{C}$-steady states. When the energies $\lambda_l + V(c_{j(l)})$ lie in a region where g is decreasing, the functional $J_{\mathcal{K}}$ in (8.17) is convex. The convexity inequality should play a role in the stability analysis.

f) Finally, as explained in the last paragraph, a more refined analysis provides a guess for possible sudden transitions from steady currents to oscillating ones as observed in experiments.

8.3.2 Algorithms

A more detailed analysis shows that the coefficients t_l are different from 0 or 1 when there are equalities of Agmon distance or equalities of energy levels. They can be thought as Lagrange multipliers, the number of undefined parameters t_l corresponding to the number of equations given by the constraint. Hence the problem of finding all the possible asymptotic steady states is well posed.

It is more natural to work with a function g which decays on $[0, +\infty)$. As mentioned above this situation can be recovered after introducing another parameter ε and considering the limit $\varepsilon \to 0$ for the family of problems associated with $g^\varepsilon(\lambda) = \chi(\varepsilon^{-1}\lambda)g(\lambda)$ with $\mathrm{supp}\,\chi \subset (0, +\infty)$, $\chi \equiv 1$ for $\lambda \geq 1$. One gets the same kind of asymptotic equation as in Theorem 8.2.2 but with an undetermined $t_l \in [0, 1]$ when $\lambda_l + V(c_l) = 0$. If one forgets the constraints on the Agmon distance, the variational formulation says that there is one unique solution for every one of the 3^N cases sorted by all the possible choices $\lambda_l + V(c_l) < 0$, $\lambda_l + V(c_l) = 0$ and $\lambda_l + V(c_l) > 0$.

Finally, it is not easy to understand the geometry of the constraints on the Agmon distances for $N > 1$. A simple way to handle it numerically is by using a penalization method. If one looks first for situations where there is no interaction between the wells, the penalization method reduces to

$$t_l = \cfrac{1}{1 + \exp\left(\cfrac{d_{(Ag)}(a,c;*)) - d_{Ag}(c,b;*)}{\nu}\right)},$$

where $\nu > 0$ is a small parameter. This parameter has to be adjusted manually: Accurate results require a small value for ν while a very small ν affects the convergence of the Newton algorithm.

8.3.3 One-well case

Here we compare our results with the results obtained by Jona-Lasinio, Presilla and Sjöstrand in [8][13] where they explain the hysteresis phenomenon without developing completely the asymptotic analysis [2]. By taking the unknown $V(c_1)$ or equivalently $\mathcal{E} = \lambda_1 + V(c_1)$ where λ_1 depends linearly on B, they derive the equation

$$\mathcal{E} = \lambda_1(B) + Cg(\mathcal{E}), \quad \text{by assuming} \quad \mathbf{c_1} - \mathbf{a} \prec\prec \mathbf{b} - \mathbf{c_1}, \tag{8.18}$$

where $C = C(a, b, c_1)$ is a positive constant determined by the positions a, c_1 and b. The number of possible steady states is 1, 2 or 3 depending on the relative position of the curves $y = x - \lambda_1(B)$ and $y = Cg(x)$ as shown below.

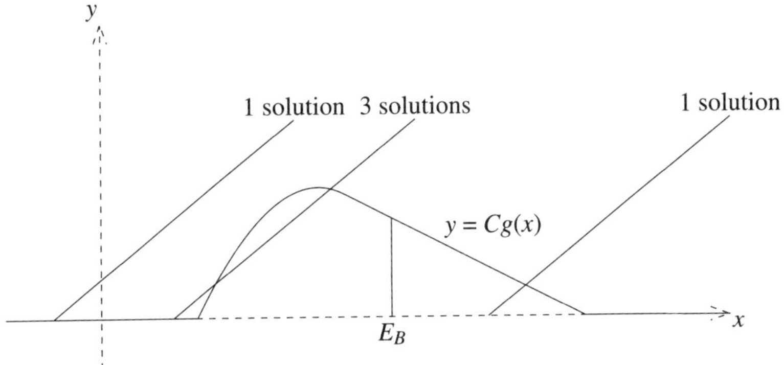

Fig. 8.2. Intersection of $y = x - \lambda_1(B)$ and $y = Cg(x)$.

Meanwhile, the equation (8.15) leads in this case to

$$\mathcal{E} = \lambda_1(B) + t_1 Cg(\mathcal{E}), \tag{8.19}$$

with $t_1 = 1$ when $d_{Ag}(a, c; \mathcal{E}, V) < d_{Ag}(c, b; \mathcal{E}, V)$, $t_1 = 0$ when $d_{Ag}(a, c; \mathcal{E}, V) > d_{Ag}(c, b; \mathcal{E}, V)$ and $t_1 \in [0, 1]$ when $d_{Ag}(a, c; \mathcal{E}, V) = d_{Ag}(c, b; \mathcal{E}, V)$. In

---

[2] Such a result was also rigorously derived by G. Perelman in unpublished notes for a simplified nonlinearity.

the Agmon distance comparison, the potential V is completely determined by $\mathcal{E}$ and B and it is rather easy to analyze the one-variable equation in $\mathcal{E}$,

$$d_{\mathrm{Ag}}(a, c; \mathcal{E}, V) = d_{\mathrm{Ag}}(c, b; \mathcal{E}, V) \quad \text{with } B \text{ given.} \tag{8.20}$$

For $b - c < c - a$: Equation (8.20) admits no solution for any $B \geq 0$ and one must take $t_1 = 0$. **For $b - c \geq c - a$:** Equation (8.20) admits a unique solution $E_B \geq 0$ and one must take $t_1 = 1$ for $\mathcal{E} < E_B$, $t_1 = 0$ for $\mathcal{E} > E_B$ and $t_1 \in [0, 1]$ for $\mathcal{E} = E_B$.

The difference between the first and the second result can be viewed graphically: For $b - c \geq c - a$ with $\theta = \frac{c-a}{b-a}$ close to $1/2$, the graph $y = g(x)$ has to be replaced by the truncated one along $x = E_B$. Hence a hysteresis phenomenon (which occurs in the presence of three possible solutions) can appear when $\theta = \frac{c-a}{b-a} \leq 1/2$ while it is impossible for $\theta = \frac{c-a}{b-a} > 1/2$. The transition from one case to the other one can be understood after checking that E_B decreases as θ (and B) increases.

The second picture shows how the treatment of the constraint on the Agmon distance via a penalization process affects the numerical results.

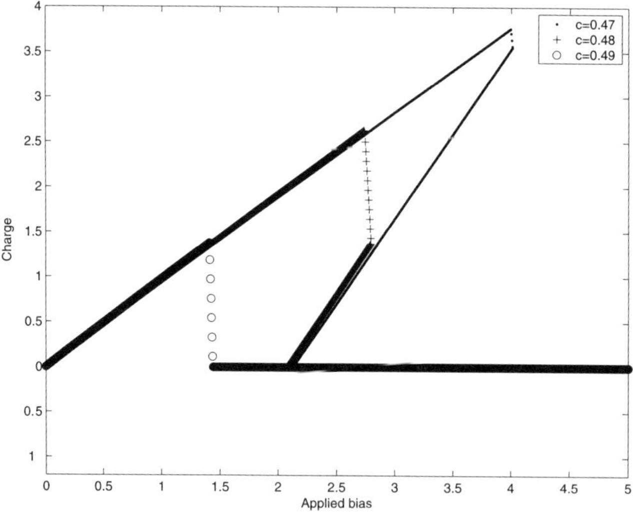

Fig. 8.3. Vanishing of the hysteresis phenomenon as $c = \theta$ approaches $1/2$

8.3.4 Situation with two wells

Even when the constraints on the Agmon distances are not active, that is in cases where the wells are closer to a than b, the bifurcation diagram can be quite complicated. The next numerical results obtained for two resonances in

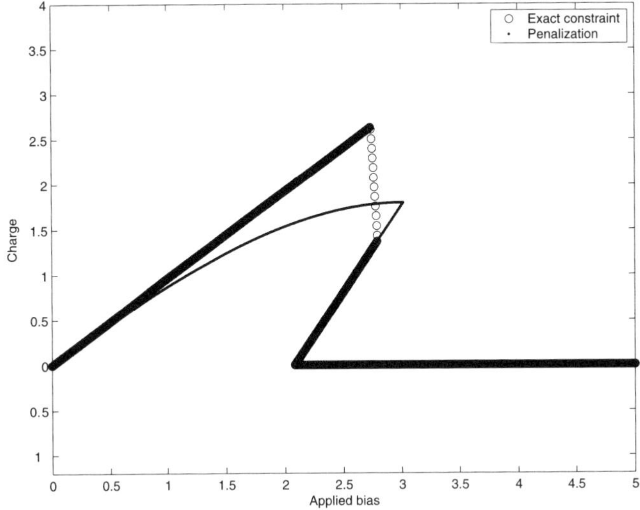

Fig. 8.4. Numerical results: Penalization versus exact constraint.

two wells (with artificial data) show the possible existence of seven steady states. Four of them are presumably stable and three of them unstable:

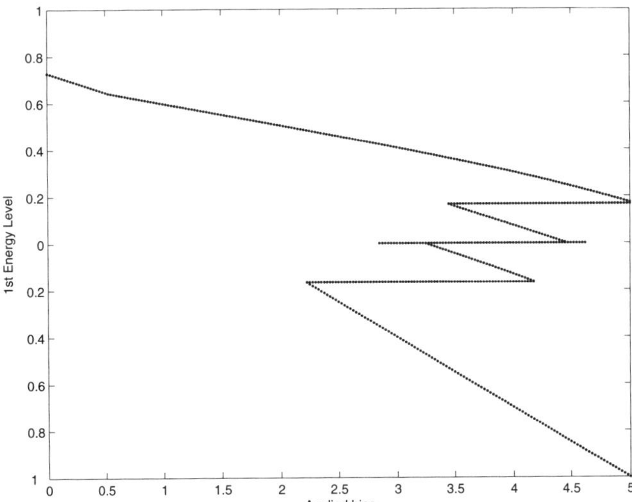

Fig. 8.5. Two wells $\theta_1 = 0.3$ and $\theta_2 = 0.45$: The energy level in the first well.

Finally let us mention the case where the first well is closer to a than b and the second well closer to b than a:

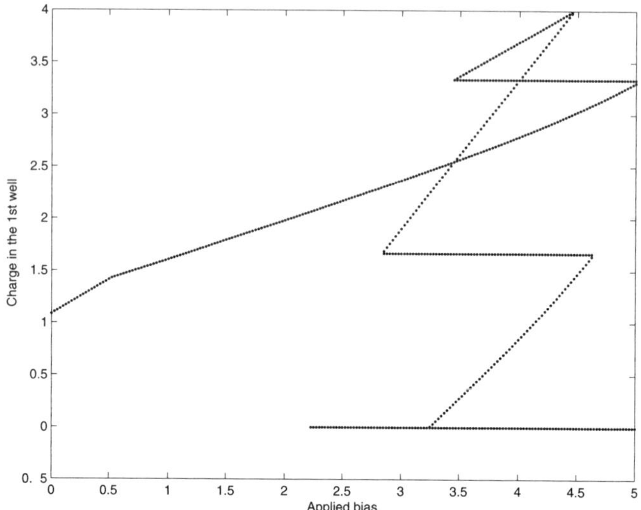

Fig. 8.6. Two wells $\theta_1 = 0.3$ and $\theta_2 = 0.45$: The charge in the first well.

$$\frac{c_1 - a}{b - a} < 1/2 \quad \text{and} \quad \frac{c_2 - a}{b - a} > 1/2\,.$$

When the energies $\mathcal{E}_1 = \lambda_1 + V(c_1)$ and $\mathcal{E}_2 = \lambda_2 + V(c_2)$ are different the only possible coefficients due to the Agmon distance comparison are $t_1 = 1$ and $t_2 = 0$ and the problem reduces to the one well case. However when $\mathcal{E}_1 = \mathcal{E}_2$, there is a possibility of having any $(t_1, t_2) \in [0, 1]^2$. A bifurcation to this case can occur quite rapidly when the applied bias B varies. Here the question is whether this new steady state is stable or if some nonlinear beating effect between the two wells occurs. This might explain experimental sudden transitions from steady currents to oscillating ones. It requires, in this framework involving quantum resonances instead of bound states, an adaptation of the recent article [6] of Grecchi, Martinez and Sacchetti about the stability of the nonlinear beating effect for ammoniac molecules coupled to a mean electromagnetic field.

References

1. N. Ben Abdallah, P. Degond, and P.A. Markowich, On a one-dimensional Schrödinger–Poisson scattering model, *Z. Angew. Math. Phys.* **48**:1 (1997), 135–155.
2. N. Ben Abdallah, On a multidimensional Schrödinger–Poisson scattering model for semiconductors, *J. Math. Phys.* **41**:7 (2000), 4241–4261.
3. L.L. Bonilla et al., Self-oscillations of domains in doped GaAs-AlAs superlattices, *Phys. Rev. B* **52**:19 (November 1995), 13761–13764.

4. C. Gérard and A. Martinez, Semiclassical asymptotics for the spectral function of long-range Schrödinger operators, *J. Funct. Anal.* **84**:1 (1989), 226–254.
5. V. Grecchi, A. Martinez, and A. Sacchetti, Double well Stark effect: Crossing and anticrossing of resonances, *Asymptotic Analysis* **13** (1996), 373–391.
6. V. Grecchi, A. Martinez, and A. Sacchetti, Destruction of the beating effect for a non-linear Schrödinger equation, *Commun. Math. Phys.* **227** (2002), 191–209.
7. B. Helffer and J. Sjöstrand, Résonances en limite semi-classique, *Mém. Soc. Math. France* **399** (1984).
8. G. Jona-Lasinio, C. Presilla, and J. Sjöstrand, On Schrödinger equations with concentrated nonlinearities, *Ann. Phys.* **240**:1, 1995.
9. W. Kirsch and B. Simon, Universal lower bound on eigenvalue splittings for one dimensional Schrödinger operators, *Commun. Math. Phys.* **97** (1985), 453–460.
10. F. Nier, Schrödinger–Poisson systems in dimension $d \leq 3$: The whole-space case, *Proc. Roy. Soc. Edinburgh* **123A** (1993), 1179–1201.
11. F. Nier, A variational formulation of Schrödinger–Poisson systems in dimension ≤ 3, *CPDE* **18**:7–8 (1993), 1125–1147.
12. F. Nier, The dynamics of some quantum open systems with short-range non-linearities, *Nonlinearity* **11** (1998), 1127–1172.
13. C. Presilla and J. Sjöstrand, Transport properties in resonant tunneling heterostructures, *J. Math. Phys.* **37**:10 (1996), 4816–4844.

9

Decoherence-induced Classical Properties in Infinite Quantum Systems

R. Olkiewicz

Institute of Theoretical Physics
University of Wrocław
Pl. Maxa Borna 9
P–50-204 Wrocław, Poland

9.1 Introduction

In recent years decoherence has been widely discussed and accepted as a mechanism responsible for the emergence of classicality in quantum measurements and the absence in the real world of Schrödinger-cat-like states [2,6,7,11,12]. Let me start with recalling the Schrödinger cat paradox. It was a hypothetical experiment proposed by Schrödinger in 1935 which can be briefly described as follows.

> A cat is penned up in a steel chamber, along with the following device: in a Geiger counter there is a tiny amount of radioactive substance, so small, that perhaps in the course of one hour one of the atoms decays, but also, with equal probability, perhaps none. If it happens, the counter tube discharges and through a relay releases a hammer which shatters a small flask of hydrocyanic acid. If one has left the entire system to itself for an hour, one would say that the cat still lives if meanwhile no atom has decayed. The first atomic decay would have poisoned it.

Therefore, after one hour, the system should be described by a quantum state which is a linear superposition over live and dead states of the cat. It is not a surprise that such an experiment caused some confusion among physicists. The standard interpretation of quantum mechanics, the so-called Copenhagen interpretation, which says that the recognition that interaction between the measuring tools and the physical system under investigation constitute an integral part of quantum phenomena, was not convincing for some physicists. They simply could not accept that quantum mechanics requires us to regard any question concerning the status of the cat as meaningless until we establish an observational relationship with it. For example in a letter to Schrödinger in 1950 Einstein wrote:

> You are the only contemporary physicist, besides Laue, who sees that
> one cannot get around the assumption of reality – if only one is honest.
> Most of them simply do not see what sort of risky game they are play-
> ing with reality – reality as something independent of what is exper-
> imentally established. Their interpretation is, however, refuted most
> elegantly by your system of radioactive atom + amplifier + charge of
> gun powder + cat in a box, in which the ψ-function of the system
> contains the cat both alive and blown to bits. Nobody really doubts
> that the presence or absence of the cat is something independent of
> the act of observation.

Our intuition tells us that after one hour the state of the cat should be a statistical mixture of live and dead states of it, the so-called classical states, with
equal probabilities. The main idea of decoherence was to get rid of such para-
doxes. It claims that classicality is an emergent property induced in quantum
open systems by their environment. Quantum interference effects for macroscopic systems are practically unobservable because superpositions of their
quantum states are effectively destroyed by the surrounding environment.
However, when people speak of classical properties induced in a quantum
system by its environment they usually mean superselection rules or pointer
states. In this talk I will try to show that new perspectives arise when we pass
from quantum systems with a finite number of degrees of freedom to quantum
systems in the thermodynamic limit.

9.2 Algebraic framework

Everybody agrees that concepts of classical and quantum physics are opposite in many aspects. Therefore, in order to demonstrate how quanta become
classical we express them in one algebraic framework. In this approach observables of any physical system are represented by self-adjoint elements of
some operator algebra $\mathcal{M}$ acting in a Hilbert space associated with the system. Genuine quantum systems are represented by factors, i.e., algebras with
a trivial center $Z(\mathcal{M}) = \mathbf{C} \cdot \mathbf{1}$, $\mathbf{1}$ stands for the identity operator, whereas
classical systems are represented by commutative algebras. Since a classical
observable by definition commutes with all other observables, it belongs to the
center of algebra $\mathcal{M}$. Hence the appearance of classical properties of a quantum system results in the emergence of an algebra with a nontrivial center,
while transition from a noncommutative to commutative algebra corresponds
to the passage from quantum to classical description of the system.

In order to study decoherence, analysis of the evolution of the reduced
density matrices obtained by tracing out the environmental variables is the
most convenient strategy. More precisely, the joint system composed of a
quantum system and its environment evolves unitarily with the Hamiltonian
H consisting of three parts:

$$H = H_S \otimes \mathbf{1}_E + \mathbf{1}_S \otimes H_E + H_I. \tag{9.1}$$

The time evolution of the reduced density matrix is then given by

$$\rho_t = \mathrm{Tr}_E(e^{-\frac{i}{\hbar}tH}(\rho_0 \otimes \omega_E)e^{\frac{i}{\hbar}tH}), \tag{9.2}$$

where Tr_E denotes the partial trace with respect to the environmental variables, and ω_E is a reference state of the environment. Alternatively, one may define the time evolution in the Heisenberg picture by

$$T_t(A) = P_E(e^{\frac{i}{\hbar}tH}(A \otimes \mathbf{1}_E)e^{-\frac{i}{\hbar}tH}), \tag{9.3}$$

where $A \in \mathcal{M}$ is an observable of the system and P_E denotes the conditional expectation onto the algebra $\mathcal{M}$ with respect to the reference state ω_E. Superoperators T_t being defined as the composition of a *-automorphism and conditional expectation satisfy in general a complicated integro-differential equation. However, for a large class of models, this evolution can be approximated by a dynamical semigroup $T_t = e^{tL}$, whose generator L is given by a Markovian master equation, see [1,8].

We are now in a position to discuss rigorously the dynamical emergence of classical observables. As was shown in [9] for each (up to some technical assumptions) Markov semigroup T_t on $\mathcal{M}$ one may associate a decomposition

$$\mathcal{M} = \mathcal{M}_1 \oplus \mathcal{M}_2 \tag{9.4}$$

such that both $\mathcal{M}_1$ and $\mathcal{M}_2$ are T_t-invariant and the following properties hold:

(i) $\mathcal{M}_1$ is a subalgebra of $\mathcal{M}$ and the evolution T_t when restricted to $\mathcal{M}_1$ is reversible, given by a one-parameter group of *-automorphisms of $\mathcal{M}_1$.
(ii) $\mathcal{M}_2$ is a linear space (closed in the norm topology) such that for any observable $B = B^* \in \mathcal{M}_2$ and any statistical state ρ of the system there is

$$\lim_{t \to \infty} \langle T_t B \rangle_\rho = 0, \tag{9.5}$$

where $\langle A \rangle_\rho$ stands for the expectation value of an observable A in state ρ.

The above result means that any observable A of the system may be written as a sum $A = A_1 + A_2$, $A_i \in \mathcal{M}_i$, $i = 1, 2$, and all expectation values of the second term A_2 are beyond experimental resolution after the decoherence time. Hence all possible outcomes of the process of decoherence can be directly expressed by the description of subalgebra $\mathcal{M}_1$ and its reversible evolution. Therefore they can be classified in the following way.

- $\mathcal{M}_1$ is noncommutative and $Z(\mathcal{M}_1) \neq \mathbf{C} \cdot \mathbf{1}$. Such a case corresponds to environment induced superselection rules [4].

- $\mathcal{M}_1$ is commutative and $T_t|_{\mathcal{M}_1}$ is trivial. In such a case we speak of environment induced pointer states [4,5].

- $\mathcal{M}_1$ is commutative and $T_t|_{\mathcal{M}_1}$ is given by a nontrivial flow on the configuration space of $\mathcal{M}_1$. In such a case we speak of environment induced continuous dynamics [4,9,10].

- $\mathcal{M}_1$ is still a factor but with significantly different properties. In such a case we say that environment changes the properties of a given system, for example it may reduce the number of degrees of freedom [3].

It is worth noting that the above classification is complete. Clearly, in discussion on emergence of classical properties the case when $\mathcal{M}_1$ is commutative is the most interesting one.

9.3 Examples

Continuous pointer states [5]. The apparatus is a semi-infinite linear array of spin-$\frac{1}{2}$ particles, fixed at positions $n = 1, 2, \ldots$ at infinite temperature. The algebra $\mathcal{M}$ of its (bounded) observables is given by the σ-weak closure of $\pi(\otimes_1^\infty M_{2\times 2})$, where π is a (faithful) GNS representation with respect to a tracial state tr on the Glimm algebra. Since there is no free evolution of the apparatus, $H_A = 0$.

The reservoir is chosen to consist of noninteracting phonons of an infinitely extended one-dimensional harmonic crystal at the inverse temperature $\beta = \frac{1}{kT}$. Since the phonons are noninteracting, their dynamics is completely determined by the energy operator

$$II_E = II_0 \otimes I - I \otimes H_0,$$

$$H_0 = d\Gamma(\omega) = \int \omega(k) a_F^*(k) a_F(k) dk,$$

where $\omega(k) = |k|$, a_F^* (a_F) are creation (annihilation) operators in the Fock representation, and $\hbar = 1$, $c = 1$. The reference state of the reservoir is taken to be a gauge-invariant quasi-free thermal state given by

$$\omega_E(a^*(f)a(g)) = \int \rho(k) \bar{g}(k) f(k) dk,$$

where

$$\rho(k) = \frac{1}{e^{\beta\omega(k)} - 1}$$

is the thermal equilibrium distribution. Clearly, ω_E is invariant with respect to the free dynamics of the environment. The Hamiltonian H of the joint system consists of the reservoir term H_E and an interacting part H_I. We assume that the coupling is linear (as in the spin-boson model), i.e., $H_I = \lambda Q \otimes \phi(g)$, where

$$Q = \pi \left(\sum_{n=1}^{\infty} \frac{1}{2^n} \sigma_n^3 \right),$$

σ_n^3 is the third Pauli matrix in the n-th site, and $\lambda > 0$ is a coupling constant. Using the so-called singular coupling limit one gets the following master equation for dynamics of the apparatus,

$$L(X) = ib[X, Q^2] + \lambda a(QXQ - \tfrac{1}{2}\{Q^2, X\}).$$

It can be easily checked that for the semigroup $T_t = e^{tL}$ the decomposition (9.4) holds.

Theorem 1. $\mathcal{M}_1$ *is a continuous commutative subalgebra of functions on the configuration space of the one-dimensional Ising model. Moreover, $T_t|_{\mathcal{M}_1}$ is trivial.*

Continuous classical dynamics [10]. Again, the system is a semi-infinite linear array of spin-$\frac{1}{2}$ particles, fixed at positions $n = 1, 2, \ldots$ at infinite temperature. The free evolution of the system is given by a σ-weakly continuous one-parameter group of automorphisms $\alpha_t : \mathcal{M} \to \mathcal{M}$ constructed in the following way. Suppose $U(\frac{k}{2^n})$, $k = 0, 1, \ldots, 2^n - 1$, is a representation of a cyclic group $\{\frac{k}{2^n}\}$, with addition modulo 1, in the space $\mathbf{C}^{2^n}$, such that

$$U(\frac{1}{2^n})(z_1, \ldots, z_{2^n}) = (z_{2^n}, z_1, \ldots, z_{2^n-1}).$$

Since it is a restriction of the standard unitary representation of the permutation group S_{2^n}, the $U(\frac{k}{2^n})$ are unitary matrices in $M_{2^n \times 2^n}$. Because there is an embedding of $M_{2^n \times 2^n}$ into $\mathcal{M}$, so they induce a discrete group of unitary automorphisms of $\mathcal{M}$,

$$\alpha_{\frac{k}{2^n}}(X) = \pi(U(\frac{k}{2^n}))X\pi(U(\frac{k}{2^n}))^*.$$

This homomorphism extends to the whole set of real numbers, yielding a group of unitary automorphisms

$$\alpha_t(X) = e^{itH}Xe^{-itH}.$$

It is clear from the construction that $\alpha_m = \mathrm{id}$, for any integer m. Suppose now that dynamics of the system is given by the master equation

$$\frac{d}{dt}X = L(X) = i[H, X] + L_D(X),$$

where the dissipative part L_D is constructed in the following way. On a subalgebra $M_{2^n \times 2^n}$ we put $L_D(A) = L_n \circ A$, where $\circ$ stands for the Hadamard (entrywise) product and L_n is a $2^n \times 2^n$ matrix given by

$$L_n = -\frac{1}{2^{n-1}}\begin{pmatrix} 0 & 1 & \cdots & 2^n - 1 \\ 1 & 0 & \cdots & 2^n - 2 \\ \cdots\cdots\cdots\cdots\cdots\cdots\cdots \\ 2^n - 1 & 2^n - 2 & \cdots & 0 \end{pmatrix}.$$

It is clear that operator L_D can be extended to a bounded and dissipative operator on the whole algebra $\mathcal{M}$ and so the operator L generates a quantum dynamical semigroup T_t on $\mathcal{M}$. Again one can check that the decomposition (9.4) holds for the semigroup T_t.

Theorem 2. $(\mathcal{M}_1, T_t)$ *is isomorphic with the classical dynamical system* (S^1, U_t), *where* U_t *is a uniform rotation of the circle* S^1.

References

1. R. Alicki and K. Lendi, *Quantum Dynamical Semigroups and Applications*, Springer-Verlag, Berlin, 1987.
2. Ph. Blanchard et al., eds., *Decoherence: Theoretical, Experimental and Conceptual Problems*, Springer-Verlag, Berlin, 2000.
3. Ph. Blanchard, P. Lugiewicz and R. Olkiewicz, *Phys. Lett. A* **314** (2003), 29.
4. Ph. Blanchard and R. Olkiewicz, *Rev. Math. Phys.* **15** (2003), 217.
5. Ph. Blanchard and R. Olkiewicz, *Phys. Rev. Lett.* **90** (2003), 010403.
6. D. Giulini et al., *Decoherence and the Appearance of a Classical World in Quantum Theory*, Springer-Verlag, Berlin, 1996.
7. E. Joos and H.D. Zeh, *Z. Phys. B* **59** (1985), 223.
8. A.J. Leggett et al., *Rev. Mod. Phys.* **59** (1987), 1.
9. P. Lugiewicz and R. Olkiewicz, *Commun. Math. Phys.* **239** (2003), 241.
10. P. Lugiewicz and R. Olkiewicz, *J. Phys. A* **35** (2002), 6695.
11. R. Omnes, *Phys. Rev A* **65** (2002), 052119.
12. W.H. Zurek, *Phys. Rev. D* **26** (1982), 1862.

Classical versus Quantum Structures: The Case of Pyramidal Molecules

C. Presilla, G. Jona-Lasinio, and C. Toninelli

Dipartimento di Fisica
Università di Roma "La Sapienza"
Piazzale Aldo Moro 2
I-00185 Roma, Italy
and
Istituto Nazionale di Fisica Nucleare
Sezione di Roma 1
I-00185 Roma, Italy
and
Istituto Nazionale per la Fisica della Materia
Unità di Roma 1 and Center for Statistical Mechanics and Complexity
I-00185 Roma, Italy

Summary. In a previous paper we proposed a model to describe a gas of pyramidal molecules interacting via dipole-dipole interactions. The interaction modifies the tunneling properties between the classical equilibrium configurations of the single molecule and, for sufficiently high pressure, the molecules become localized in these classical configurations. The model explains quantitatively the shift to zero-frequency of the inversion line observed upon increase of the pressure in a gas of ammonia or deuterated ammonia. Here we analyze further the model especially with respect to stability questions.

10.1 Introduction

The behavior of gases of pyramidal molecules, i.e., molecules of the kind XY_3 like ammonia NH_3, has been the object of investigations since the early developments of quantum mechanics [11]. In recent times the problem has been discussed again in several papers [4, 12, 17, 7, 13, 14] from a stationary point of view while in [8, 9, 10] a dynamical approach has been attempted. For a short historical sketch of the issues involved we refer to [4, 17, 13].

In [13] we have constructed a simplified mean-field model of a gas of pyramidal molecules which allows a direct comparison with experimental data. Our model predicts, for sufficiently high inter-molecular interactions, the presence of two degenerate ground states corresponding to the different localizations of the molecules. This transition to localized states gives a reasonable explanation of the experimental results [1, 2, 3]. In particular, it describes quantitatively, without free parameters, the shift to zero-frequency of the inversion line of NH_3 and ND_3 on increasing the pressure.

In the present paper we first reconsider our model from the standpoint of stationary many-body theory, clarifying the meaning of the mean-field energy levels. We then analyze the mean field states with respect to the energetic stability. The conclusions agree with those found in [10] via a dynamical analysis of the same type of model to which dissipation is added.

10.2 The model

We model the gas as a set of two-level quantum systems, that mimic the inversion degree of freedom of an isolated molecule, mutually interacting via the dipole-dipole electric force.

The Hamiltonian for the single isolated molecule is assumed of the form $-\frac{\Delta E}{2}\sigma^x$, where σ^x is the Pauli matrix with symmetric and antisymmetric delocalized tunneling eigenstates φ_+ and φ_-,

$$\sigma^x = \begin{pmatrix} 0 & 1 \\ 1 & 0 \end{pmatrix} \qquad \varphi_+ = \frac{1}{\sqrt{2}}\begin{pmatrix} 1 \\ 1 \end{pmatrix} \qquad \varphi_- = \frac{1}{\sqrt{2}}\begin{pmatrix} 1 \\ -1 \end{pmatrix}. \tag{10.1}$$

Since the rotational degrees of freedom of the single pyramidal molecule are faster than the inversion ones, on the time scales of the inversion dynamics set by ΔE the molecules feel an effective attraction arising from the angle averaging of the dipole-dipole interaction at the temperature of the experiment [15]. The localizing effect of the dipole-dipole interaction between two molecules i and j can be represented by an interaction term of the form $-g_{ij}\sigma_i^z\sigma_j^z$, with $g_{ij} > 0$, where σ^z is the Pauli matrix with left and right localized eigenstates φ_L and φ_R,

$$\sigma^z = \begin{pmatrix} 1 & 0 \\ 0 & -1 \end{pmatrix} \qquad \varphi_L = \begin{pmatrix} 1 \\ 0 \end{pmatrix} \qquad \varphi_R = \begin{pmatrix} 0 \\ 1 \end{pmatrix}. \tag{10.2}$$

The Hamiltonian for N interacting molecules then reads

$$H = -\frac{\Delta E}{2}\sum_{i=1}^{N} 1_1 \otimes 1_2 \otimes \cdots \otimes \sigma_i^x \otimes \cdots \otimes 1_N$$

$$-\sum_{i=1}^{N}\sum_{j=i+1}^{N} g_{ij}\, 1_1 \otimes \cdots \otimes \sigma_i^z \otimes \cdots \otimes \sigma_j^z \otimes \cdots \otimes 1_N. \tag{10.3}$$

For a gas of moderate density, we approximate the behavior of the $N \gg 1$ molecules with the mean-field Hamiltonian

$$h[\psi] = -\frac{\Delta E}{2}\sigma^x - G\langle\psi, \sigma^z\psi\rangle\sigma^z, \tag{10.4}$$

where ψ is the single-molecule state ($\langle\psi,\psi\rangle = 1$) to be determined self-consistently by solving the nonlinear eigenvalue problem associated to (10.4). The parameter G represents the dipole interaction energy of a single molecule with the rest of the gas. This must be identified with a sum over all possible molecular distances and all possible dipole orientations calculated with the Boltzmann factor at temperature T. Assuming that the equation of state for an ideal gas applies, we find [13]

$$G = \frac{4\pi}{9}\left(\frac{T_0}{T}\right)^2 Pd^3, \tag{10.5}$$

where $T_0 = \mu^2/(4\pi\varepsilon_0\varepsilon_r d^3 k_B)$, ε_0 and ε_r being the vacuum and relative dielectric constants, d the molecular collision diameter and μ the molecular electric dipole moment. Note that, at fixed temperature, the mean-field interaction constant G increases linearly with the gas pressure P.

10.3 Molecular states

The nonlinear eigenvalue problem associated to (10.4), namely

$$h[\psi_\mu]\psi_\mu = \mu\psi_\mu \qquad \langle\psi_\mu,\psi_\mu\rangle = 1, \tag{10.6}$$

has different solutions depending on the value of the ratio $G/\Delta E$. If $G/\Delta E < \frac{1}{2}$, we have only two solutions corresponding to the delocalized eigenstates of an isolated molecule

$$\psi_{\mu_1} = \varphi_+ \qquad\qquad \mu_1 = -\Delta E/2, \tag{10.7}$$
$$\psi_{\mu_2} = \varphi_- \qquad\qquad \mu_2 = +\Delta E/2. \tag{10.8}$$

If $G/\Delta E > \frac{1}{2}$, there appear also two new solutions,

$$\psi_{\mu_3} = \sqrt{\frac{1}{2}+\frac{\Delta E}{4G}}\,\varphi_+ + \sqrt{\frac{1}{2}-\frac{\Delta E}{4G}}\,\varphi_- \qquad \mu_3 = -G, \tag{10.9}$$

$$\psi_{\mu_4} = \sqrt{\frac{1}{2}+\frac{\Delta E}{4G}}\,\varphi_+ - \sqrt{\frac{1}{2}-\frac{\Delta E}{4G}}\,\varphi_- \qquad \mu_4 = -G, \tag{10.10}$$

which in the limit $G \gg \Delta E$ approach the localized states φ_L and φ_R, respectively. Solutions (10.9) and (10.10) are termed chiral in the sense that $\psi_{\mu_4} = \sigma^x \psi_{\mu_3}$.

The states ψ_μ determined above, are the stationary solutions $\psi(t) = \exp(i\mu t/\hbar)\psi_\mu$ of the time-dependent nonlinear Schrödinger equation

$$i\hbar\frac{\partial}{\partial t}\psi(t) = h[\psi]\psi(t)\,. \tag{10.11}$$

122   C. Presilla, G. Jona-Lasinio, and C. Toninelli

The generic state $\psi(t)$ solution of this equation has an associated conserved energy given by

$$\mathcal{E}[\psi] = -\frac{\Delta E}{2}\langle\psi,\sigma^x\psi\rangle - \frac{G}{2}\langle\psi,\sigma^z\psi\rangle^2 . \tag{10.12}$$

The value of this functional calculated at the stationary solutions (10.7-10.10) provides the corresponding single-molecule energies $e_i = \mathcal{E}[\psi_{\mu_i}]$,

$$\begin{aligned}
e_1 &= -\Delta E/2, \\
e_2 &= +\Delta E/2, \\
e_3 = e_4 &= -\frac{\Delta E}{2} - \frac{1}{2G}\left(\frac{\Delta E}{2} - G\right)^2 .
\end{aligned} \tag{10.13}$$

These energies are plotted in Fig. 10.1 as a function of the ratio $G/\Delta E$. The state effectively assumed by the molecules in the gas will be that with the minimal energy, namely the symmetric delocalized state ψ_{μ_1} for $G/\Delta E < \frac{1}{2}$ or one of the two degenerate chiral states for $G/\Delta E > \frac{1}{2}$.

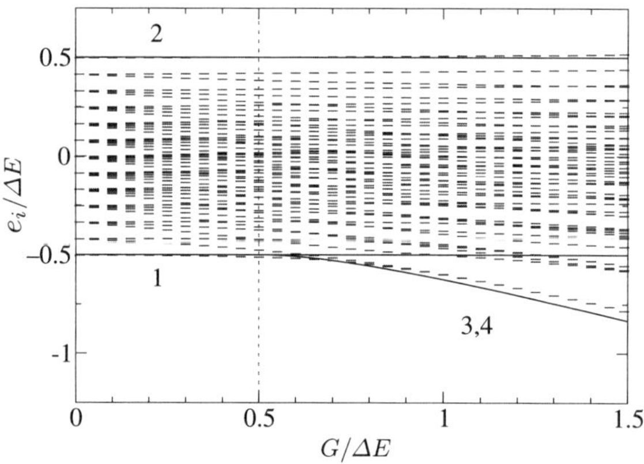

Fig. 10.1. Single-molecule energies e_i (solid lines) of the four stationary states ψ_{μ_i}, $i = 1, 2, 3, 4$, as a function of the ratio $G/\Delta E$. The dashed lines are the eigenvalues, divided by N, of the Hamiltonian (10.3) with $g_{ij} = G/N$ and $N = 12$.

The above results imply a bifurcation of the mean-field ground state at a critical interaction strength $G = \Delta E/2$. According to Eq. (10.5), this transition can be obtained for a given molecular species by increasing the gas pressure above the critical value

$$P_{\mathrm{cr}} = \frac{9}{8\pi}P_0\left(\frac{T}{T_0}\right)^2 , \tag{10.14}$$

where $P_0 = \Delta E/d^3$. In Table 10.1 we report the values of T_0 and P_0 calculated for different pyramidal molecules.

	ΔE (cm^{-1})	μ (Debye)	d (Å)	T_0 (Kelvin)	P_0 (atm)
NH_3	0.81	1.47	4.32	193.4	1.97
ND_3	0.053	1.47	4.32	193.4	0.13
PH_3	3.34×10^{-14}	0.57	–	29.1	8.11×10^{-14}
AsH_3	2.65×10^{-18}	0.22	–	4.3	6.44×10^{-18}

Table 10.1. Measured energy splitting ΔE, collision diameter d, and electric dipole moment μ, for different pyramidal molecules as taken from [16, 5]. In the fourth and fifth columns we report the temperature T_0 and the pressure P_0 evaluated as described in the text. In the case of PH_3 and AsH_3 the collision diameter, not available, is assumed equal to that measured for NH_3 and ND_3. We used $\varepsilon_r = 1$.

10.4 Inversion line

When a gas of pyramidal molecules which are in the delocalized ground state is exposed to an electromagnetic radiation of angular frequency $\omega_0 \sim \Delta E/\hbar$, some molecules can be excited from the state φ_+ to the state φ_-. For a non-interacting gas this would imply the presence in the absorption or emission spectrum of an inversion line of frequency $\bar{\nu} = \Delta E/h$. Due to the attractive dipole-dipole interaction, the value of $h\bar{\nu}$ evaluated as the energy gap between the many-body first excited level and the ground state is decreased with respect to the noninteracting case by an amount of the order of G. As shown in Fig. (10.2), the value of the inversion line frequency is actually a function of the number N of molecules and in the limit $N \gg 1$ approaches the mean field value [13]

$$\bar{\nu} = \frac{\Delta E}{h} \left(1 - \frac{2G}{\Delta E} \right)^{\frac{1}{2}}. \tag{10.15}$$

According to (10.15), the inversion line is obtained only in the range $0 \leq G \leq \Delta E/2$ and its frequency vanishes at $G = \Delta E/2$.

In [13] we have compared the mean field theoretical prediction for the inversion line with the spectroscopic data available for ammonia [1, 2] and deuterated ammonia [3]. In these experiments the absorption coefficient of a cell containing NH_3 or ND_3 gas at room temperature was measured at different pressures, i.e., according to (10.5) at different values of the interaction strength G. The measured frequency $\bar{\nu}$ decreases by increasing P and vanishes for pressures greater than a critical value. This behavior is very well accounted for by the the mean field prediction (10.15). In particular, the critical pressure evaluated according to Eq. (10.14) with no free parameters, at $T = 300$ K is $P_{\text{cr}} = 1.695$ atm for NH_3 and $P_{\text{cr}} = 0.111$ atm for ND_3 in very good agreement with the experimental data.

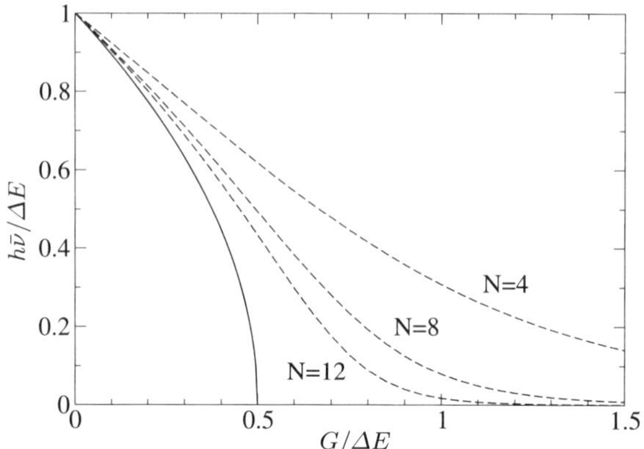

Fig. 10.2. Inversion line frequency as a function of the ratio $G/\Delta E$ in the mean field model (solid line) and obtained from the Hamiltonian (10.3) with $g_{ij} = G/N$ and $N = 4$, 8, and 12.

10.5 Energetic stability of the molecular states

In order to discuss the energetic stability of the mean field molecular states found in Section 10.3 we introduce the free energy $\mathcal{F}[\psi] = \mathcal{E}[\psi] - \mu\langle\psi,\psi\rangle$. The stationary solutions of Eq. (10.11) then can be viewed as the critical points of the Hamiltonian dynamical system

$$i\hbar\frac{\partial}{\partial t}\begin{pmatrix}\psi \\ \psi^*\end{pmatrix} = \begin{pmatrix}0 & 1 \\ -1 & 0\end{pmatrix}\begin{pmatrix}\frac{\delta\mathcal{F}[\psi]}{\delta\psi} \\ \frac{\delta\mathcal{F}[\psi]}{\delta\psi^*}\end{pmatrix}. \tag{10.16}$$

Under the effect of a perturbation which dissipates energy, a stationary state ψ_μ will remain stable only if $\mathcal{F}[\psi_\mu]$ is a minimum. Therefore we are interested in exploring the nature of the extremal values $\mathcal{F}[\psi_\mu]$ of the free energy functional. In general, this can be done in terms of the eigenvalues and the eigenvectors of the linearization matrix associated to the dynamical system (10.16) as explained in [6] for a Gross–Pitaevskii equation. Here, due to the simplicity of the model, we can provide a more direct analysis.

For a variation of the stationary solution $\psi_\mu \to \psi_\mu + \delta\phi$, up to the second order in $\delta\phi$, we have

$$\mathcal{F}[\psi_\mu + \delta\phi] = \mathcal{F}[\psi_\mu] + \delta^2\mathcal{F}[\psi_\mu, \delta\phi], \tag{10.17}$$

where

$$\delta^2\mathcal{F}[\psi_\mu, \delta\phi] = -\frac{\Delta E}{2}\langle\delta\phi, \sigma^x\delta\phi\rangle - \mu\langle\delta\phi, \delta\phi\rangle$$

$$-G\left(\langle\psi_\mu, \sigma^z\psi_\mu\rangle\langle\delta\phi, \sigma^z\delta\phi\rangle + \langle\psi_\mu, \sigma^z\delta\phi\rangle\langle\delta\phi, \sigma^z\psi_\mu\rangle\right.$$

$$\left.+\frac{1}{2}\langle\psi_\mu, \sigma^z\delta\phi\rangle^2 + \frac{1}{2}\langle\delta\phi, \sigma^z\psi_\mu\rangle^2\right). \tag{10.18}$$

The variation $\delta\phi$ can be taken in the most general form

$$\delta\phi = ae^{i\theta_a}\varphi_+ + be^{i\theta_b}\varphi_-, \tag{10.19}$$

where $\varphi_\pm$ are the delocalized tunneling eigenstates (10.1) and a, b, θ_a, and θ_b arbitrary real parameters with the constraint that $\langle\psi_\mu + \delta\phi, \psi_\mu + \delta\phi\rangle = 1 + \mathcal{O}(\delta\phi^2)$. By writing

$$\psi_\mu = a_\mu\varphi_+ + b_\mu\varphi_-, \tag{10.20}$$

with the real coefficients a_μ and b_μ deduced by Eqs. (10.7–10.10), the above constraint implies

$$a_\mu a \cos\theta_a + b_\mu b \cos\theta_b = 0. \tag{10.21}$$

The second variation of the free energy evaluated at the four stationary solutions ψ_{μ_i}, $i = 1, 2, 3, 4$, with the condition (10.21) gives

$$\delta^2\mathcal{F}[\psi_{\mu_1}] = G\,2b^2\left(\frac{\Delta E}{2G} - \cos^2\theta_b\right), \tag{10.22}$$

$$\delta^2\mathcal{F}[\psi_{\mu_2}] = G\,2a^2\left(-\frac{\Delta E}{2G} - \cos^2\theta_a\right), \tag{10.23}$$

$$\delta^2\mathcal{F}[\psi_{\mu_k}] = G\left[2a^2\left(1 + \frac{\Delta E}{2G}\right)\cos^2\theta_a + 2b^2\left(1 - \frac{\Delta E}{2G}\right)\cos^2\theta_b\right.$$
$$\left. + \left(a\sqrt{1 - \frac{\Delta E}{2G}}\sin\theta_a \mp b\sqrt{1 + \frac{\Delta E}{2G}}\sin\theta_b\right)^2\right], \tag{10.24}$$

where $k = 3, 4$ and the signs $\mp$ refer respectively to $k = 3$ and $k = 4$. We see that, for the state ψ_{μ_1}, the variation $\delta^2\mathcal{F}$ is always positive for $G < \frac{\Delta E}{2}$ and can be negative for $G > \frac{\Delta E}{2}$. The variation $\delta^2\mathcal{F}$ is always negative in the case of ψ_{μ_2}. For the states ψ_{μ_3} and ψ_{μ_4}, which exist only for $G > \frac{\Delta E}{2}$, the variation $\delta^2\mathcal{F}$ is always positive. We conclude that the free energy has a single minimum in correspondence of the delocalized state ψ_{μ_1} when $G < \frac{\Delta E}{2}$, and two degenerate minima in correspondence of the chiral states ψ_{μ_3} and ψ_{μ_4} when $G > \frac{\Delta E}{2}$. The energetic stability analysis is summarized in Table 10.2. Note that our results coincide with those reported in [10] where a standard linear stability analysis is performed for the same model considered here to which an explicit norm-conserving dissipation is added.

10.6 Conclusions

The specific prediction of our model for the critical pressure P_{cr} in terms of the electric dipole μ of the molecule, its size d, the splitting ΔE and the temperature T of the gas, successfully verified in the case of ammonia, could

	ψ_{μ_1}	ψ_{μ_2}	ψ_{μ_3}	ψ_{μ_4}
$\mathcal{F}[\psi_\mu]$	$-\Delta E$	$+\Delta E$	$\frac{G}{2}\left[1-\left(\frac{\Delta E}{2G}\right)^2\right]$	$\frac{G}{2}\left[1-\left(\frac{\Delta E}{2G}\right)^2\right]$
$G < \frac{\Delta E}{2}$	$\delta^2\mathcal{F}>0$ minimum	$\delta^2\mathcal{F}<0$ maximum		
$G > \frac{\Delta E}{2}$	$\delta^2\mathcal{F}\gtrless 0$ saddle point	$\delta^2\mathcal{F}<0$ maximum	$\delta^2\mathcal{F}>0$ minimum	$\delta^2\mathcal{F}>0$ minimum

Table 10.2. Value of the free energy $\mathcal{F}[\psi_\mu]$ and sign of its second variation at the four extrema ψ_{μ_i}, $i = 1, 2, 3, 4$.

be experimentally tested also for other pyramidal gases for which Eqs. (10.14) and (10.15) predict the scaling law

$$\frac{\bar{\nu}_{XY_3}(P)}{\bar{\nu}_{XY_3}(0)} = \frac{\bar{\nu}_{X'Y_3'}(\gamma P)}{\bar{\nu}_{X'Y_3'}(0)}, \tag{10.25}$$

where $\gamma = P_{\mathrm{cr}\,X'Y_3'} / P_{\mathrm{cr}\,XY_3}$.

Our model applies not only to molecules XY_3 but also to their substituted derivatives $XYWZ$. An important difference between the two cases is that for XY_3 the localized states can be obtained one from the other either by rotation or by space inversion, while for $XYWZ$ they can be connected only by space inversion. This implies that $XYWZ$ molecules at a pressure greater than the critical value are chiral and therefore optically active. The measurement of the optical activity of pyramidal gases for $P > P_{\mathrm{cr}}$ would allow a direct verification of this prediction.

References

1. B. Bleaney and J.H. Loubster, *Nature* **161** (1948), 522.
2. B. Bleaney and J.H. Loubster, *Proc. Phys. Soc. (London)* **A63** (1950), 483.
3. G. Birnbaum and A. Maryott, *Phys. Rev.* **92** (1953), 270.
4. P. Claverie and G. Jona-Lasinio, *Phys. Rev. A* **33** (1986), 2245.
5. *CRC Handbook of Chemistry and Physics*, 76th ed., edited by D.R. Lide, CRC Press, New York, 1995.
6. R. D'Agosta and C. Presilla, *Phys. Rev. A* **65** (2002), 043609.
7. V. Grecchi and A. Martinez, *Comm. Math. Phys.* **166** (1995), 533.
8. V. Grecchi, A. Martinez and A. Sacchetti, *Comm. Math. Phys.* **227** (2002), 191.
9. V. Grecchi and A. Sacchetti, *Phys. Rev. A* **61** (2000), 052106.
10. V. Grecchi and A. Sacchetti, *J. Phys. A: Math. Gen.* **37** (2004), 3527; http://arxiv.org/abs/math-ph/0210062.
11. F. Hund, *Z. Phys.* **43** (1927), 805.
12. G. Jona-Lasinio and P. Claverie, *Prog. Theor. Phys. Suppl.* **86** (1986), 54.
13. G. Jona-Lasinio, C. Presilla, and C. Toninelli, *Phys. Rev. Lett.* **88** (2002), 123001.

14. G. Jona-Lasinio, C. Presilla, and C. Toninelli, in *Mathematical Physics in Mathematics and Physics. Quantum and Operator Algebraic aspects*, Vol. 30, Fields Institute Communications, 2001, p. 207.
15. W.H. Keesom, *Phys. Z.* **22** (1921), 129.
16. C.H. Townes and A.L. Schawlow, *Microwave Spectroscopy*, Mc Graw-Hill, New York, 1955.
17. A.S. Wightman, *Il nuovo Cimento* **110 B** (1995), 751.

11

On the Quantum Boltzmann Equation

M. Pulvirenti

Dipartimento di Matematica
Università di Roma 'La Sapienza'
P. le Aldo Moro, 5
I–00185 Roma, Italy

Summary. In this contribution I describe the problem of deriving a Boltzmann equation for a system of N interacting quantum particles under suitable scaling limits. From a rigorous viewpoint, the problem is still open and only partial results are available, even for short times. The present report is based on a systematic collaboration with D. Benedetto, F. Castella and R. Esposito: possible mistakes and inconsistencies are however the responsibility of the author.

11.1 The problem

A large quantum system of identical interacting particles can be often conveniently described in terms of a Boltzmann equation. This description is physically meaningful only in suitable regimes, namely when the number of particles is large and the interaction is moderate (weak-coupling limit), or when the particle gas is rarefied (low-density limit).

In this contribution I try to establish the problem of a logically well founded and mathematically rigorous transition from the usual description of quantum mechanics in terms of the Schrödinger equation and the more manageable kinetic picture as given by the nonlinear Boltzmann equation associated to the system at hand. This is done by means of an appropriate scaling limit referring to the physical situation we are dealing with.

It is probably useful to note that such a transition is delicate and technically difficult because it treats time-reversible systems as irreversible ones, as largely argued for classical systems where a rigorous derivation of the classical Boltzmann equation in the low-density regime was obtained by Lanford in 1975 for short times (see [L]) and in Ref. [IP] globally in time, but for special situations (see also [CIP] for additional comments).

We finally remark that, in contrast with the quantum case, classical systems in the weak-coupling limit are described by a diffusive (in velocity) kinetic equation, the so-called Landau–Fokker–Planck equation (see [S1]). Thus the applicability of the Boltzmann equation is larger for quantum systems (where kinetic descriptions, besides those for dilute gases, including dense weakly interacting systems, as for example the electron gas in semiconductors) than for classical ones.

One pragmatic way to introduce the quantum Boltzmann equation (see e.g., [CC] and [UU]) is to solve the scattering problem in quantum mechanics and then to replace, in the classical Boltzmann equation, the classical cross section with the quantum one, taking into account the statistics when it is the case.

A more logically founded approach is to derive an evolution equation for the Wigner transform of a quantum state associated to the particle system. Working on this equation, one can hope to recover, at the quantum level, the same physical arguments as those used at the classical level to obtain propagation of chaos and a suitable kinetic description for the one-particle distribution function. This is the strategy we illustrate here.

We consider an N-particle quantum system in $\mathbb{R}^3$ and assume the mass of the particles, as well as $\hbar$, to be 1. The interaction is described by a two-body potential ϕ so that the potential energy is:

$$U(x_1 \ldots x_N) = \sum_{i<j} \phi(x_i - x_j). \tag{11.1}$$

The Schrödinger equation reads:

$$i\partial_t \Psi(X_N, t) = \frac{1}{2} \Delta_N \Psi(X_N, t) + U(X_N)\Psi(X_N, t) \tag{11.2}$$

where $\Delta_N = \sum_{i=1}^{N} \Delta_i$, Δ_i is the Laplacian with respect to the x_i variables, and X_N is a shorthand notation for $x_1 \ldots x_N$.

We rescale the equation according to the hyperbolic space-time scaling

$$x \to \varepsilon x \, , \qquad t \to \varepsilon t \tag{11.3}$$

and simultaneously we rescale also the potential $\phi \to \sqrt{\varepsilon}\phi$. Hence the resulting equation is

$$i\varepsilon \partial_t \Psi^\varepsilon(X_N, t) = \frac{\varepsilon^2}{2} \Delta_N \Psi^\varepsilon(X_N, t) + U_\varepsilon(X_N)\Psi^\varepsilon(X_N, t), \tag{11.4}$$

where

$$U_\varepsilon(x_1 \ldots x_N) = \sum_{i<j} \phi_\varepsilon(x_i - x_j) \tag{11.5}$$

and

$$\phi_\varepsilon = \sqrt{\varepsilon}\phi(\frac{x}{\varepsilon}). \tag{11.6}$$

Note that $\Psi^\varepsilon(X_N, t)$ is fully determined by Eq.(1.4) and the initial datum which will be specified later on.

We want to analyze the limit $\varepsilon \to 0$ in the above equations, while keeping

$$N = \varepsilon^{-3} . \tag{11.7}$$

This kind of limit is usually called a weak-coupling limit. Another possible scaling to be considered is the low-density limit. In this case ϕ is unscaled but $N = O(\varepsilon^{-2})$. In the classical context this is nothing but the Boltzmann–Grad limit (see for example [CIP]). We refer the interested reader to [S2], which discusses and makes clear the two scaling limits and the expected kinetic equations.

We now introduce the Wigner function:

$$W^N(X_N, V_N) = \left(\frac{1}{2\pi}\right)^{3N} \int dY_N \; e^{iY_N \cdot V_N} \overline{\Psi}^\varepsilon(X_N + \frac{\varepsilon}{2} Y_N)\Psi^\varepsilon(X_N - \frac{\varepsilon}{2} Y_N).$$

$$\tag{11.8}$$

A standard computation yields:

$$(\partial_t + V_N \cdot \nabla_N)W^N(X_N, V_N) = \frac{1}{\sqrt{\varepsilon}}\left(T_N^\varepsilon W^N\right)(X_N, V_N) \tag{11.9}$$

where $V_N \cdot \nabla_N = \sum_{i=1}^{N} v_i \cdot \nabla_{x_i}$ and $(\partial_t + V_N \cdot \nabla_N)$ is the usual free-stream operator. Also, we have introduced

$$(T_N^\varepsilon W^N)(X_N, V_N) = \sum_{0 < k < \ell \leq N} (T_{k,\ell}^\varepsilon W^N)(X_N, V_N), \tag{11.10}$$

with

$$(T_{k,\ell}^\varepsilon W^N)(X_N, V_N) = \frac{1}{i}(\frac{1}{2\pi})^{3N} \int dY_N \int dV_N' \; e^{iY_N \cdot (V_N - V_N')} W^N(X_N, V_N')$$

$$\times \left[\phi\left(\frac{x_k - x_\ell}{\varepsilon} - \frac{1}{2}(y_k - y_\ell)\right) - \phi\left(\frac{x_k - x_\ell}{\varepsilon} + \frac{1}{2}(y_k - y_\ell)\right)\right] . \tag{11.11}$$

In other words,

$$(T_{k,\ell}^\varepsilon W^N)(X_N, V_N) = -i \sum_{\sigma=\pm 1} \sigma \int \frac{dh}{(2\pi)^3}\hat{\phi}(h)e^{i\frac{h}{\varepsilon}(x_k - x_\ell)}$$

$$\tag{11.12}$$

$$\times W^N\left(x_1, v_1, \ldots, x_k, v_k - \frac{\sigma h}{2}, \ldots, x_\ell, v_\ell + \frac{\sigma h}{2}, \ldots, x_N, v_N\right) .$$

The operator $T_{k,\ell}^\varepsilon$ describes the "collision" of particle k with particle ℓ, and the total operator T_N^ε takes all possible "collisions" into account. Here and below, $\hat{f} = (\mathcal{F}_\S\{)(\langle)$ denotes the Fourier transform of f.

We now introduce the Wigner transform of the partial traces according to the formula, for $j = 1, \ldots, N - 1$:

$$f_j^N(X_j, V_j) = \int dx_{j+1} \cdots \int dx_N \int dv_{j+1} \cdots \int dv_N$$
$$\times W^N(X_j, x_{j+1} \ldots x_N; V_j, v_{j+1} \ldots v_N). \tag{11.13}$$

Obviously, we set $f_N^N = W^N$.

From now on we shall suppose that, due to the fact that the particles are identical, the objects which we have introduced ($\Psi^\varepsilon, W^N, f_j^N$) are all symmetric in the exchange of particles.

Proceeding as in the derivation of the BBKGY hierarchy for classical systems (see e.g., [CIP]), we readily arrive at the following hierarchy of equations (for $1 \le j \le N$):

$$(\partial_t + \sum_{k=1}^{j} v_k \cdot \nabla_k)f_j^N = \frac{1}{\sqrt{\varepsilon}}T_j^\varepsilon f_j + \frac{N-j}{\sqrt{\varepsilon}}C_{j+1}^\varepsilon f_{j+1}^N. \tag{11.14}$$

The operator C_{j+1}^ε is defined as

$$C_{j+1}^\varepsilon = \sum_{k=1}^{j} C_{k,j+1}^\varepsilon , \tag{11.15}$$

and

$$C_{k,j+1}^\varepsilon f_{j+1}(x_1 \ldots x_j; v_1 \ldots v_j)$$
$$= -i \sum_{\sigma=\pm 1} \sigma \int \frac{dh}{(2\pi)^3} \int dx_{j+1} \int dv_{j+1} \, \hat{\phi}(h) \tag{11.16}$$
$$\times e^{i\frac{h}{\varepsilon}(x_k - x_{j+1})} f_{j+1}(x_1, x_2, \ldots, x_{j+1}, v_1, \ldots, v_k - \sigma\frac{h}{2}, \ldots, v_{j+1} + \sigma\frac{h}{2}).$$

The operator $C_{k,j+1}^\varepsilon$ describes the "collision" of particle k, belonging to the j-particle subsystem, with a particle outside the subsystem, conventionally denoted by the number $j+1$ (this numbering uses the fact that all particles are identical). The total operator C_{j+1}^ε takes into account all such collisions. As usual (see e.g., [CIP]), Eq. (11.14) shows that the dynamics of the j-particle subsystem is governed by three effects: the free-stream operator, the collisions "inside" the subsystem (the T term), and the collisions with particles "outside" the subsystem (the C term).

We finally fix the initial value $\{f_j^0\}_{j=1}^N$ of the solution $\{f_j^N(t)\}_{j1}^N$ assuming that $\{f_j^0\}_{j=1}^N$ is factorized, that is, for all $j = 1, N$,

$$f_j^0 = f_0^{\otimes j}, \tag{11.17}$$

where f_0 is a one-particle Wigner function which we assume also to be a probability distribution. We recall that a quantum state, whose Wigner transform is a general positive f_0, is not in general a wave function but rather a density matrix. As a consequence the evolution equation we have to use is not the Schrödinger equation (1.4) but rather the Heisenberg equation for the density matrix. In both cases the corresponding Wigner equation is Eq. (11.9) or, equivalently, Eq. (11.14).

We expect that in the limit $\varepsilon \to 0$ the one-particle distribution function f_1^N converges to the solution of a suitable Boltzmann equation f which we are going to introduce.

Let $f(t)$, $t \in [0, t_0)$ be the solution of the "classical" Boltzmann equation

$$(\partial_t + v \cdot \nabla_x)f = Q(f, f) \tag{11.18}$$

where

$$Q(f, f)(x, v) = \int dv_1 \int d\omega B(\omega, |v - v_1|)$$
$$\times (f(x, v')f(x, v_1') - f(x, v)f(x, v_1)) \tag{11.19}$$

and v' and v_1' are the outgoing velocities after a collision with impact parameter $\omega \in S^2$ and incoming velocities v and v_1. Explicitly,

$$v' = v - (v - v_1) \cdot \omega \quad \omega,$$
$$v_1' = v_1 - (v - v_1) \cdot \omega \quad \omega.$$

Finally B is the cross-section and depends on the interaction potential. In the weak-coupling limit the collisions take place with a small energy and in a scale distance of order ε. Therefore the cross section must be computed via the quantum rules and at low energy. In other words it is expected to be the Born approximation, namely:

$$B(\omega, w) = \frac{1}{8\pi^2}|\omega \cdot w| \, |\hat{\phi}(\omega \, (\omega \cdot w))|^2. \tag{11.20}$$

What we have illustrated so far, called Problem 1 in the sequel, is the weak-coupling limit for particles without statistics or obeying the Maxwell–Boltzmann (M–B) statistics.

A more interesting problem (Problem 2 in the sequel) is however to consider particles obeying the Fermi–Dirac (F–D) or Bose–Einstein (B–E) statistics. In this case, although the dynamics is that introduced above, we cannot assume the independence property (1.17), because the statistics yields correlations even at time zero. The most uncorrelated state one can introduce not violating the F–D or B–E statistics, are called quasi-free. The program of showing the Boltzmann equation for such system should pass through the characterization of the quasi-free states in terms of the Wigner function and hence replacing condition (1.17) by a more appropriate independence property taking into account the statistics. The one-particle distribution function is then expected to converge to the solution of the Boltzmann equation (1.18) with:

$$Q(f,f)(x,v) = \int\int dv_1 d\omega B_\theta(\omega, |v - v_1|) \times \qquad (11.21)$$

$$[f(x,v')f(x,v_1')(1 - \theta f(x,v)f(x,v_1)) - f(x,v)f(x,v_1)(1 - \theta f(x,v')f(x,v_1'))]$$

where $\theta = +1$ and $\theta = -1$ according to the B–E and F–D statistics respectively and B_θ is the symmetrized or antisymmetrized cross-section.

Finally Problems 3 is concerned with the low-density limit. In this case, due to the rarefaction limit, we expect that all the statistics asymptotically coincides. The expected Boltzmann equation is still Eq.(1.18), with the collision term given by Eq.(1.19). Indeed the particles are too rare to make effective the statistical correlations . However the collision takes place at large energies so that the cross-section should be the full one and not that at first order in the Born expansion.

In the next section I will briefly discuss the very few results concerning the above three problems.

11.2 Detecting the leading terms

As regards Problem 1, one can try to handle the hierarchy (1.14) as for the Boltzmann–Grad limit for classical systems, namely to study the asymptotic behavior of the solution expressed in terms of the series expansion for $1 \leq j \leq N$, obtained by iterating the Duhamel formula,

$$f_j^N(t) = \sum_{n=0}^{N-j} \frac{(N-j)\dots(N-j-n)}{(\sqrt{\varepsilon})^n} \int_0^t dt_1 \cdots \int_0^{t_{n-1}} dt_n \, S_{\text{int}}^\varepsilon(t - t_1)C_{j+1}^\varepsilon$$

$$\times S_{\text{int}}^\varepsilon(t_1 - t_2)C_{j+2}^\varepsilon \dots S_{\text{int}}^\varepsilon(t_{n-1} - t_n)C_{j+n}^\varepsilon S_{\text{int}}^\varepsilon(t_n)f_{j+n}^0. \qquad (11.22)$$

Here $S_{\text{int}}^\varepsilon(t)f_j$ is the j-particle interacting flow, namely the solution to the initial value problem:

$$\begin{cases} (\partial_t + V_j \cdot \nabla_j)S_{\text{int}}^\varepsilon(t)f_j = \frac{1}{\sqrt{\varepsilon}}T_j^\varepsilon S_{\text{int}}^\varepsilon(t)f_j, \\ \\ S_{\text{int}}^\varepsilon(0)f_j = f_j. \end{cases} \qquad (11.23)$$

If we expand $S_{\text{int}}^\varepsilon(t)$ as a perturbation of the free flow $S(t)$ defined as

$$(S(t)f_j)(X_j, V_j) = f_j(X_j - V_j t, V_j), \qquad (11.24)$$

we find

$$S_{\text{int}}^\varepsilon(t)f_j = S(t)f_j + \sum_{m \geq 0} \frac{1}{(\sqrt{\varepsilon})^m} \int_0^t d\tau_1 \int_0^{\tau_1} d\tau_2 \cdots \int_0^{\tau_{m-1}} d\tau_m \, S(t - \tau_1)T_j^\varepsilon$$

$$S(\tau_1 - \tau_2)T_j^\varepsilon \ldots S(\tau_{m-1} - \tau_m)T_j^\varepsilon S(\tau_m)f_j. \tag{11.25}$$

Inserting (2.4) into (2.1), the resulting series contains a huge number of terms. However, we expect that many of these contributions are negligible in the limit. For instance if we analyze up to the second order terms (in the potential) the full expansion, we find the following four terms:

$$\mathcal{I}_0 = S(t)f_j^0 \,, \tag{11.26}$$

$$\mathcal{I}_1 = \frac{N-j}{\sqrt{\varepsilon}} \int_0^t dt_1 \ S(t-t_1)C_{j+1}^\varepsilon S(t_1)f_{j+1}^0 \,, \tag{11.27}$$

$$\mathcal{I}_2 = \frac{1}{\sqrt{\varepsilon}} \int_0^t d\tau_1 \ S(t-\tau_1)T_j^\varepsilon S(\tau_1)f_j^0 \,, \tag{11.28}$$

$$\mathcal{I}_3 = \frac{N-j}{e} \int_0^t d\tau_1 \int_0^{\tau_1} dt_1 \ S(t-\tau_1)T_j^\varepsilon S(\tau_1-t_1)C_{j+1}^\varepsilon S(t_1)f_{j+1}^0 \,, \tag{11.29}$$

$$\mathcal{I}_4 = \frac{N-j}{\sqrt{\varepsilon}} \int_0^t dt_1 \int_0^{t_1} d\tau_1 \ S(t-t_1)C_{j+1}^\varepsilon S(t_1-\tau_1)T_{j+1}^\varepsilon S(\tau_1)f_{j+1}^0$$

$$= \sum_{r=1}^{j} \sum_{1 \le s < \ell \le j+1} \mathcal{I}_4^{r,\ell,s} \,, \tag{11.30}$$

where

$$\mathcal{I}_4^{r,\ell,s} = \frac{N-j}{\sqrt{\varepsilon}} \int_0^t dt_1 \int_0^{t_1} d\tau_1 \ S(t-t_1)C_{r,j+1}^\varepsilon S(t_1-\tau_1)T_{\ell,s}^\varepsilon S(\tau_1)f_{j+1}^0. \tag{11.31}$$

It is possible to show (see [BCEP]) that $\mathcal{I}_i$, $i = 1, 2, 3$ are negligible in the limit $\varepsilon \to 0$ for oscillation or cancellations. Also, all the contributions to $\mathcal{I}_4$ except those for $r = \ell$ and $s = j + 1$, corresponding to a collision/recollision event, are equally vanishing. As matter of fact, the only $O(1)$ term is $\mathcal{I}_4^{\ell,\ell,j+1}$ which corresponds to a creation-recollision event. Let us compute this term for $\ell = j = 1$. It is:

$$\mathcal{I}_4^{1,1,2} =$$

$$-\frac{N-j}{\varepsilon} \sum_{\sigma,\sigma'=\pm 1} \sigma\sigma' \int_0^t dt_1 \int_0^{t_1} d\tau_1 \int dx_2 \int dv_2 \int \frac{dh}{(2\pi)^3} \int \frac{dk}{(2\pi)^3}$$

$$\times \hat{\phi}(h)\hat{\phi}(k)e^{i\frac{h}{\varepsilon}\cdot\left(x_1-x_2-v_1(t-t_1)\right)} e^{i\frac{k}{\varepsilon}\cdot\left(x_1-x_2-(v_1-v_2)(t-t_1)-(v_1-v_2-\sigma_1 h)(t_1-\tau_1)\right)}$$

$$\times f_2^0\left(x_1 - v_1 t + \frac{\sigma h}{2}t_1 + \frac{\sigma'k}{2}\tau_1, x_2 - v_2 t - \frac{\sigma h}{2}t_1 - \frac{\sigma'k}{2}\tau_1; \right. \tag{11.32}$$

$$\left. v_1 - \frac{\sigma h}{2} - \frac{\sigma'k}{2}, v_2 + \frac{\sigma h}{2} + \frac{\sigma'k}{2}\right).$$

Making the following change of variables:

$$t_1 - \tau_1 = \varepsilon s_1 \,, \qquad (\text{ i.e. } \ \tau_1 = t_1 - \varepsilon s_1) \,,$$
$$\xi = (h + k)/\varepsilon \,. \tag{11.33}$$

we arrive at

$$\mathcal{I}_4^{1,1,2} = \tag{11.34}$$

$$-(N-j)\varepsilon^3 \sum_{\sigma,\sigma'=\pm 1} \sigma\sigma' \int_0^t dt_1 \int_0^{t_1/\varepsilon} ds_1 \int dx_2 \int dv_2 \int \frac{d\xi}{(2\pi)^3} \int \frac{dk}{(2\pi)^3}$$

$$\times \hat{\phi}(-k+\varepsilon\xi_1)\hat{\phi}(k)e^{i\xi\cdot\left(x_1-x_2-v_1(t-t_1)\right)}e^{-is_1k\cdot(v_1-v_2-\sigma(-k+\varepsilon\xi))}f_{j+1}^0(\dots)\,.$$

In the limit $\varepsilon \to 0$ we recover the asymptotics,

$$\mathcal{I}_4^{1,1,2} \sim -\sum_{\sigma,\sigma'=\pm 1} \sigma\sigma' \int_0^t dt_1 \int_0^{+\infty} ds_1 \int dv_2 \int \frac{dk}{(2\pi)^3}|\hat{\phi}(k)|^2 e^{-is_1k\cdot(v_1-v_2+\sigma k)}$$

$$\times f_2^0\Big(x_1 - v_1 t - \frac{(\sigma-\sigma')k}{2}t_1, x_2 - v_2 t + \frac{(\sigma-\sigma')k}{2}t_1; \tag{11.35}$$
$$v_1 + \frac{(\sigma-\sigma')k}{2}, v_2 - \frac{(\sigma-\sigma')k}{2}\Big)$$

which can be justified from a rigorous viewpoint (see [BCEP]). Moreover

$$\mathrm{Re}\int_0^\infty ds_1 \, e^{-is_1k\cdot(v_1-v_2+\sigma k)} = -\pi\delta(k\cdot(v_1-v_2+\sigma k))\,. \tag{11.36}$$

Using formula (2.15) we realize that the contribution $\sigma = -\sigma'$ gives rise to the gain term:

$$\int_0^t dt_1 \int dv_2 \int d\omega B(\omega,|v_1-v_2|) \tag{11.37}$$
$$f_2^0(x_1 - v_1(t-t_1) - v_1't_1, x_2 - v_2(t-t_1) - v_2't_1; v_1', v_2')\,.$$

Notice that k is the momentum transferred in the collision and that the δ in (2.15) expresses nothing else than the conservation of the energy in the collision. The momentum conservation is automatically satisfied.

The term $\sigma = \sigma'$ yields the loss part:

$$\int_0^t dt_1 \int dv_2 \int d\omega B(\omega,|v_1-v_2|) \tag{11.38}$$
$$\times f_2^0(x_1 - v_1 t, x_2 - v_2(t-t); v_1, v_2)\,.$$

We finally remark that the imaginary part of the time integral in the left hand side of (2.15) does not give any contribution because of cancellations.

All these steps have been rigorously proved in [BCEP] and this shows, in a sense, that our quantum system agrees with the Boltzmann evolution up to

the second order. This is far from conclusive since there are examples (see e.g.
the pathologies of the Broadwell model in [CIP]) in which the agreement fails
at the fourth order. However it is possible to show that the subseries formed by
all the contributions' creation-recollision terms is indeed convergent, for short
times, and approaching, in the limit $\varepsilon \to 0$, the corresponding series obtained
by solving the Boltzmann equation. We note also that a rigorous proof of
the term-by-term convergence is still missing. Even more difficult is to find
a bound of the full series, thus a mathematical justification of the quantum
Boltzmann equation is a still open, challenging and difficult problem.

The situation is more complicated when considering the case of bosons
and Fermions (Problem 2). The statistics involves the structure of the states
and a complete factorization is not compatible with B–E or F–D statistics.
Indeed systems of independent particles are called quasi-free and have reduced
density matrices satisfying the property

$$\rho_j(x_1 \dots x_j; y_1 \dots y_j) = \sum_{\pi \in \mathcal{P}_j} \prod_{i=1}^{j} \rho(x_i; y_{\pi(i)}) \tag{11.39}$$

where $\rho(x,y)$ is the kernel of a one-particle density matrix and $\mathcal{P}_j$ is the group
of the permutations of j elements.

Writing down the explicit expression of the Wigner transforms of quasi-
free states, we can investigate the asymptotics of the series expansion of the
solution up to the second order in the same sense explained above for the
M–B statistics. We actually recover Eq. (11.21) for a suitable B_θ. This work
is in progress.

We mention that a similar analysis, using commutator expansion in the
framework of the second quantization formalism, has been performed in [HL]
(following [H]) in the case of the Van Hove limit for lattice systems (that is the
same as for the weak-coupling limit without rescaling the distances). For more
recent formal results in this direction, but in the context of the weak-coupling
limit, see [ESY].

We finally observe that in the low-density regime (Problem 3) the number of vanishing terms is much less. Indeed while for the weak-coupling the
only $O(1)$ terms are those of the form CT, for the low-density also the terms
$CT \dots T$ are $O(1)$. Actually they must be resummed to recover the full cross-
section, making convergent (under suitable smallness assumptions on the potential) the Born series. Also this is work in progress.

References

[BCEP]   D. Benedetto, F. Castella, R. Esposito, and M.Pulvirenti, Some considerations on the derivation of the nonlinear quantum Boltzmann equation,
Math. Phys. Archive, University of Texas (2003), 3–19, and *J. Stat. Phys.*
(to appear).

[CC]   S. Chapman and T. G. Cowling, *The Mathematical Theory of Non-uniform Gases*, Cambridge Univ. Press, Cambridge, England, 1970.

[CIP]   C. Cercignani, R. Illner, and M. Pulvirenti, The mathematical theory of dilute gases, *Applied Mathematical Sciences*, Vol. 106, Springer-Verlag, New York, 1994.

[ESY]   L. Erdös, M. Salmhofer, and H.-T. Yau, On the quantum Boltzmann equation, *Math. Phys. Archive, University of Texas* (2003), 3–19.

[H]   M.N. Hugenholtz, Derivation of the Boltzmann equation for a Fermi gas, *J. Stat. Phys* **32**(1983), 231–254.

[HL]   N.T. Ho and L.J. Landau, Fermi gas in a lattice in the van Hove limit, *J. Stat. Phys.* **87** (1997), 821–845.

[IP]   R. Illner and M. Pulvirenti, Global validity of the Boltzmann equation for a two-dimensional rare gas in the vacuum, *Comm. Math. Phys.* **105** (1986), 189–203; Erratum and improved result, *Comm. Math. Phys.* **121** (1989), 143–146.

[L]   O. Lanford III, Time evolution of large classical systems, *Lecture Notes in Physics*, Vol. 38, E.J. Moser, ed., Springer-Verlag, 1975, pp. 1–111.

[S1]   H. Spohn, *Large Scale Dynamics of Interacting Particles*, Springer-Verlag, 1991.

[S2]   H. Spohn, Quantum kinetic equations, in *On Three Levels: Micro-meso and Macro Approaches in Physics*, M.Fannes, C.Maes, A Verbeure, eds., *Nato Series B: Physics* **324** (1994), 1–10.

[UU]   E.A. Uehling and G.E. Uhlembeck, Transport phenomena in Einstein–Bose and Fermi–Dirac gases. I, *Phys. Rev.* **43** (1933), 552–561.

12

Remarks on Time-dependent Schrödinger Equations, Bound States, and Coherent States

D. Robert

Laboratoire de Mathématiques Jean Leray
CNRS-UMR N° 6629
Université de Nantes
F–44322 Nantes-Cedex 03, France

Summary. It has been well known from the beginning of quantum theory that there exist deep connections between the time evolution of a classical Hamiltonian system and the bound states for the Schrödinger equation, in particular in the semiclassical régime. These connections are well understood for integrable systems (Bohr–Sommerfeld quantization rules). But for more intricate systems (like classically chaotic Hamiltonian) the mathematical analysis of the bound states is much more difficult and there are few rigorous mathematical results. In this paper our goal is to revisit some of these results and to show that they can be proven, and sometimes improved, by using essentially two technics: the Wigner–Weyl calculus and the propagation of observables on one side, the propagation of coherent states on the other side. We want to emphasize that in our approach we get rather explicit estimates in terms of classical dynamics.

The main ideas explained here, in particular the use of coherent states, are the results of several years of collaboration with Monique Combescure.

12.1 Introduction

One of the most important problems in quantum mechanics is to compute matrix elements $A_{jk}(\hbar) = \langle \hat{A}\varphi_j, \varphi_k \rangle$ (transition amplitudes), where $\hat{A}$ is an observable and $\{\varphi_j\}$ is an orthonormal system of normalized bound states of a given quantum Hamiltonian $\hat{H}$. $\hat{H}$ is supposed to be a self-adjoint operator in the Hilbert space in $L^2(\mathbb{R}^d)$ for a system with d degrees of freedom.

In the semiclassical régime considered here, $\hat{H}$ is obtained as the $\hbar$-Weyl quantization of a classical Hamiltonian H. We have $\hat{H}\varphi_j = E_j\varphi_j$, where E_j is the eigenenergy of φ_j.

The diagonal matrix elements $A_{jj}(\hbar)$ are clearly related with trace formulas. Assume for simplicity in a first step that $\hat{H} = -\frac{\hbar^2}{2}\Delta + V$ where V is a smooth confining electric potential, i.e., $\lim_{|q|\to+\infty} V(q) = +\infty$ (like a polynomial). $A(q, p)$ can be any smooth classical observable on the phase space with polynomial behaviour in (q, p) at ∞.

Let $\{\varphi_j\}_{j\geq 0}$ be an orthonormal basis of bounded states. So we have

$$\mathrm{Tr}(\hat{A}f(\hat{H})) = \sum_{j\geq 0} f(E_j)A_{jj}(\hbar) \,.$$

Consider, for example, the Gibbs states at temperature $T = \beta^{-1}$, $\hat{G}(\beta) = \mathrm{e}^{-\beta\hat{H}}/\mathrm{Tr}(\mathrm{e}^{-\beta\hat{H}})$. By standard application of Weyl–Wigner calculus, it is easily proved that

$$\mathrm{Tr}\left(\hat{A}\mathrm{e}^{-\beta\hat{H}}\right) \asymp (2\pi\hbar)^{-d}\sum_{j\geq 0} g_{\beta,j}\hbar^{2j} \ \ \text{with} \ \ g_{\beta,0} = \int_{\mathbb{R}^{2d}} A(z)\mathrm{e}^{-\beta H(z)}dz \quad (12.1)$$

and

$$\lim_{\hbar\to 0}\mathrm{Tr}(\hat{A}\hat{G}(\beta)) = \langle A\rangle_\beta \ \ \text{where} \ \ \langle A\rangle_\beta = \frac{\int_{\mathbb{R}^{2d}} A(z)\mathrm{e}^{-\beta H(z)}dz}{\int_{\mathbb{R}^{2d}} \mathrm{e}^{-\beta H(z)}dz} \quad (12.2)$$

and $H(z) = \frac{p^2}{2} + V(q)$, $z = (q, p)$.

These two asymptotics give a rough average behaviour for the energies E_j and the matrix elements $\hat{A}_{jj}(\hbar)$. To get more accurate information, as it is well known, we need to work in a small window in the energy spectrum, $E_j \in [E - \delta, E + \delta]$ where E is a fixed classical energy and $\delta > 0$ is as small as possible. But doing that, time-dependent phenomena occur, as it is expected from the time-energy uncertainty principle. More precisely, the classical dynamics of the Hamiltonian H enter the game when δ is of the same order as $\hbar$. This is transparent with the Gutzwiller trace formula which displays a semiclassical asymptotic for

$$\Xi_{\rho,A}(\hbar) = \sum_{j\geq 0}\rho\left(\frac{E_j - E}{\hbar}\right)A_{jj}(\hbar) \quad (12.3)$$

where the Fourier transform $\tilde{\rho}$ of ρ is a smooth function with bounded support. Only periodic trajectories of energy E of the classical Hamiltonian contribute in the asymptotic expansion of $\Xi_{\rho,A}(\hbar)$. This is also true for the average of an observable in the Gibbs state if the temperature is low, of the same order of the Planck constant $\hbar$ (see [6]).

A closely related problem is to estimate the counting function of the eigenenergies in a fixed real interval $I = [E', E]$. Let us denote by $N_I(\hbar)$ the number of bound states of $\hat{H}$ in I. Under generic assumptions (E', E are not critical for V) we have the Weyl law

$$N_I(\hbar) = (2\pi\hbar)^{-d}\int_{H(z)\in I} dz + O(\hbar^{1-d}) \,.$$

As we shall see later, a more difficult problem is to find a second term and to estimate the error. Some properties of the classical flow at large time are also needed because periodic trajectories give oscillatory contributions. This is already obvious for the harmonic oscillator ($V(q) = q^2$).

When the classical dynamics is chaotic (ergodic) on the energy shell $\Sigma_E := H^{-1}(E)$ (E noncritical), we shall also discuss the quantum ergodic theorem, whose meaning is the following. Let be $I_\hbar = [E - \delta_\hbar, E + \delta_\hbar]$ shrinking to E in a suitable way. Then, except for a negligible set of eigenenergies in $I_\hbar$, we have

$$\lim_{\hbar \to 0, \; E_j(\hbar) \in I_\hbar} A_{jj}(\hbar) = \int_{\Sigma_E} A(z) d\nu_E(z) \tag{12.4}$$

where $d\bar\nu_E$ is the normalized Liouville measure on Σ_E.

The links between time-dependent and time-independent phenomena appear also clearly in the following question. It is conjectured that the behaviour of $\hat{A}_{jk}(\hbar)$ resembles a random matrix model. (see for example[28, 29]) For classically chaotic systems, Wilkinson [28] conjectures that the matrix elements $A_{jk}(\hbar)$ are independent, Gaussian, with mean zero when $j \neq k$. The last statement is supported by results proved in ([4]). The variance introduced by Wilkinson is

$$\mathcal{V}_{(f,g)}(\hbar, E, \tau) =$$
$$\sum_{[E_j(\hbar), E_k(\hbar) \in I_\hbar]} |A_{jk}(\hbar)|^2 f_\hbar \left(E - \tfrac{1}{2}(E_j(\hbar) + E_k(\hbar)) \right) . g\left(\tau - \omega_{jk}(\hbar) \right) \tag{12.5}$$

where E is inside the interval $I_\hbar$, $\omega_{jk}(\hbar) = \frac{E_j(\hbar) - E_k(\hbar)}{\hbar}$, f, g are Gaussian regularizations of the Dirac δ_0 distribution. $f_\hbar(u) := \frac{1}{\hbar} f(\frac{u}{\hbar})$ with $f(u) = \frac{1}{\sigma_1 \sqrt{2\pi}} e^{u^2/2\sigma_1^2}$ and $g(u) = \frac{1}{\sigma_2 \sqrt{2\pi}} e^{u^2/2\sigma_2^2}$, $\sigma_1, \sigma_2 > 0$.

This variance has a nice semiclassical limit. If the Fourier transform $\tilde{f}$ of f has a small support around zero, then

$$\lim_{\hbar \to 0} (2\pi\hbar)^d \mathcal{V}_{(f,g)}(\hbar, E, \tau) = \tilde{f}(0) \int \tilde{g}(t) e^{it\tau} C_A(E, t) dt$$

where $\tilde{f}$ (Fourier transform of f) has a small support around zero and $C_A(E, t)$ is the classical autocorrelation function

$$C_A(E, t) = \int_{\Sigma_E} A(z) A(\Phi_H^t(z)) d\nu_E(z) .$$

If the support of f contains some nonzero periods of the classical flow, we have oscillatory terms, as in the Gutzwiller trace formula as we shall see later.

It seems much more difficult to prove a similar result for Gaussian test functions. The main difficulty comes from f: we need some control of the large periods of the classical flow which are not well known up to now. Concerning g

one can use recent results obtained in [1] where accurate estimates for longtime propagation of observables are proved.

Our main goal in this paper is to revisit some already known connections between time-dependent and time-independent properties for Schrödinger equations by adding when possible explicit estimates. We will give some ideas about proofs, showing that they can be deduced essentially from two basic long-time dependent accurate results: propagation of observables and propagation of coherent states. This last technique will be used instead of the BKW method to avoid the well-known caustic problem. Moreover this way we can get time-dependent estimates up to the Ehrenfest time (of order $|\log \hbar|$ in the chaotic case, $\hbar^{-\kappa}$, $\forall \kappa < 1/8$, in the integrable case). These time-dependent estimates are useful, for example, to improve the remainder estimate in the Weyl formula with two terms or to control the speed of convergence in the quantum ergodic theorem.

The content of the paper is the following.

We first recall from [1] the results concerning propagation of observables. Then we recall from [6, 25] the main facts concerning coherent states and their propagation by the time-dependent Schrödinger equation.

We will explain how to apply these tools to different spectral problems. We discuss the trace Gutzwiller formula and the main steps to prove it according to the method used in [3]. Then we apply this to the Weyl law with two terms and remainder estimate. We also discuss the proof of the semiclassical expansion for the Wilkinson variance.

In the last section we explain some results concerning diagonal and non-diagonal matrix elements $A_{jk}(\hbar)$, in particular the quantum ergodic theorem and related topics.

12.2 Propagation of observables

Let us denote by $X = \mathbb{R}^d$ the configuration space of a classical mechanical system with d degrees of freedom. The corresponding phase space Z is identified with $\mathbb{R}^{2d}$ equipped with the symplectic form σ defined by

$$\sigma(z, z') = Jz \cdot z' \tag{12.6}$$

where $\cdot$ is the Euclidean scalar product and J is the $2d \times 2d$ matrix

$$J = \begin{pmatrix} 0 & \mathbb{I}_d \\ -\mathbb{I}_d & 0 \end{pmatrix}. \tag{12.7}$$

A generic point in Z is denoted by z and its coordinates by (q, p) where $q, p \in \mathbb{R}^d$.

A classical Hamiltonian is a smooth real function $H : Z \to \mathbb{R}$. Our basic example will be $H(q, p) = \frac{|p|^2}{2m} + V(q)$ $(m > 0)$, where $|p|^2 = p \cdot p$.

The motion of the classical system is determined by the system of Hamilton's equations

$$\frac{dq_t}{dt} = \frac{\partial H}{\partial p}(q_t, p_t), \quad \frac{dp_t}{dt} = -\frac{\partial H}{\partial q}(q_t, p_t) \; ; (q_0, p_0) = (q, p). \tag{12.8}$$

The equations (12.8) generate a flow Φ_H^t on the phase space Z, defined by $\Phi_H^t(q, p) = (q_t, p_t)$. Let us consider a classical observable A, i.e., A a smooth real-valued function defined on Z. The time evolution of A is given by

$$\frac{d}{dt} A(\Phi^t(z)) = \{H, A\}(\Phi^t(z)), \tag{12.9}$$

where $\{H, A\}$ is the Poisson bracket defined by

$$\{H, A\} = \partial_q H \cdot \partial_p A - \partial_p H \cdot \partial_q A. \tag{12.10}$$

Here we have used the notation $\partial_q = \frac{\partial}{\partial q}$. By Weyl quantization of H and A, we get quantum observables $\hat{H}$ and $\hat{A}$ in $L^2(X)$ with $\hat{H}$ self-adjoint. So we can define the one-parameter group of unitary operators $U(t) = \exp\left(-\frac{it}{\hbar}\hat{H}\right)$. The quantum time evolution of the observable $\hat{A}$ is given by $\hat{A}(t) = U(-t)\hat{A}U(t)$ which satisfies the Heisenberg–von Neumann equation:

$$\frac{d\hat{A}(t)}{dt} = \frac{i}{\hbar}[\hat{H}, \hat{A}], \tag{12.11}$$

where $[K, B] = KB - BK$ is the commutator of K, B. Let us recall the $\hbar$-Weyl quantization formula, for $A \in \mathcal{S}(Z)$ (the space of Schwartz functions) and for $\psi \in \mathcal{S}(X)$, we have:

$$Op_\hbar^w A\psi(x) = \hat{A}\psi(x) = (2\pi\hbar)^{-d} \iint_Z A\left(\frac{x+y}{2}, p\right) e^{i\hbar^{-1}(x-y)\cdot p}\psi(y)dydp. \tag{12.12}$$

This definition can be extended to the following classes of observables.

Definition 1.
(i) A weight function on the phase space Z is a positive continuous function μ on Z such that there exist $C > 0$, $M \geq 0$ such that for every $z, z' \in Z$,

$$\mu(z) \leq C(1 + |z - z'|)^M \mu(z').$$

(ii) $A \in \mathcal{O}(\mu)$, where μ is a weight, if and only if $Z \xrightarrow{A} \mathbb{C}$ is C^∞ in Z and for every multi-index $\gamma \in \mathbb{N}^{2d}$ there exists $C_\gamma > 0$ such that

$$|\partial_z^\gamma A(z)| \leq C_\gamma \mu(z), \; \forall z \in Z.$$

(iii) We say that A is a semiclassical observable of weight μ if there exist $\hbar_0 > 0$ and a sequence $A_j \in \mathcal{O}(\mu)$, $j \in \mathbb{N}$, such that for every $N \in \mathbb{N}$ and every $\gamma \in \mathbb{N}^{2d}$ there exists $C_N > 0$ such that for all $\hbar \in]0, \hbar_0[$ we have

$$\sup_Z \mu^{-1}(z) \left| \partial_z^\gamma \left(A(\hbar, z) - \sum_{0 \leq j \leq N} \hbar^j A_j(z) \right) \right| \leq C_N \hbar^{N+1}. \tag{12.13}$$

A_0 is called the principal symbol, A_1 the sub-principal symbol of $\hat{A}$. The set of semiclassical observables of weight μ is denoted by $\mathcal{O}_{sc}(\mu)$. Its range by the $\hbar$-Weyl quantization is denoted $\widehat{\mathcal{O}}_{sc}(\mu)$.

If $\mu(z) = \mu_m(z) = (1 + |z|)^m$, $m \in \mathbb{R}$, we say that the observable is of weight m.

Notation : For any A and A_j satisfying (12.13), we will write : $A(\hbar) \asymp \sum_{j \geq 0} \hbar^j A_j$ in $\mathcal{O}_{sc}(\mu)$.

Let us now recall from [1] a statement for the propagation of observables which will be useful for applications to bound states.

Assume that $H \in \mathcal{O}_{sc}(\mu)$. Let Ω be a bounded open set in the phase space Z. We assume that the closure $\overline{\Omega}$ of Ω is invariant by the flow $\Phi^t := \Phi^t_{H_0}$ ($\forall t \in \mathbb{R}$) and that there exists an increasing function s from $]0, \infty[$ in $[1, +\infty[$ satisfying $s(T) \geq T$ and such that the following estimates are satisfied:

$$\sup_{z \in \Omega, |t| \leq T} \left| \partial_z^\gamma \Phi^t(z) \right| \leq C_\gamma s(T)^{|\gamma|} \tag{12.14}$$

where C_γ depends only on $\gamma \in \mathbb{N}^{2d}$. Then we have:

Theorem 1. *For every smooth observable A, with compact support in Ω, $\widehat{A}(t)$ is a semiclassical observable of weight $-\infty$, with semiclassical symbol supported in Ω, such that*

$$A(t, \hbar) \asymp \sum_{j \geq 0} A_j(t) \hbar^j$$

where

$$A_0(t, z) = A(\Phi^t(z)), \quad A_1(t, z) = \int_0^t \{A(\Phi^\tau), H_1\}(\Phi^{t-\tau}(z)) d\tau \tag{12.15}$$

and for $j \geq 2$, by induction,

$$A_j(t, z) = \sum_{\substack{|(\alpha,\beta)| + k + \ell = j+1 \\ 0 \leq \ell \leq j-1}} C(\alpha, \beta) \int_0^t [(\partial_p^\alpha \partial_q^\beta H_k)(\partial_q^\alpha \partial_p^\beta A_\ell)(\tau)](\Phi^{t-\tau}(z)) d\tau, \tag{12.16}$$

with $C(\alpha, \beta) = \dfrac{(-1)^{|\beta|} - (-1)^{|\alpha|}}{\alpha! \beta! 2^{|\alpha| + |\beta|}} i^{-1 - |(\alpha,\beta)|}.$ \hfill (12.17)

Furthermore we have the following estimates in L^2 operator-norm of the remainder term:

$$\|\hat{A}(t) - \sum_{0 \leq j \leq N} \hbar^j \hat{A}_j(t)\|_{L^2} \leq C_N \hbar^{N+1}(1 + |t|)^{N+1} s(|t|)^{(2N+\epsilon_d)|t|} \qquad (12.18)$$

where C_N is independent of t and $\hbar$, ϵ_d is a universal constant ($\epsilon_d = 5d + 10$).

Remark 1. Using a classical result on ODE (Gronwall inequality), we always have estimates with exponential time: $s(T) = e^{\Lambda T}$ for some $\Lambda > 0$. If the classical system is integrable, nonsingular in $\overline{\Omega}$, then we can choose $s(T) = 1 + T$ ([1]). In particular the semiclassical régime is still valid in time intervals $[-T_\hbar, T_\hbar]$ where $T_\hbar = \frac{1-\varepsilon}{2\Lambda} |\log \hbar|$ for the general case and $T_\hbar = \hbar^{\varepsilon-1/3}$ in the integrable case, for arbitrary $\varepsilon > 0$.

Remark 2. If the expansion of H in $\hbar$ is even (in particular if H is "classical" : $H = H_0$) then the $\hbar$-expansion of $A(t)$ is even and in the remainder estimate (12.18) the term $2N$ in the exponent becomes $3N/2$. This kind of improvement may be useful for some applications, as one can see in [9].

12.3 Propagation of Gaussian coherent states

Let us consider in this section the time-dependent Schrödinger equation

$$i\hbar \frac{\partial \psi_{z,t}}{\partial t} = \widehat{H}\psi_{z,t}, \quad \psi_{z,0} = \psi_z, \qquad (12.19)$$

where $\hat{H}$ is a semiclassical self-adjoint Hamiltonian of weight $m \in \mathbb{R}$ and ψ_z is a coherent state maximized at $z \in Z$. So we have

$$\psi_{z,t} = U(t)\psi_z, \text{ with } U(t) = e^{(-it/\hbar)\hat{H}} . \qquad (12.20)$$

A well-known method to construct asymptotic solutions of (12.19) is the WKB expansion. One of the main difficulties of WKB methods comes from the occurring of caustics so that the shape of WKB approximations changes dramatically when time increases (caustics may appear at times of order 1 in the Planck scale $\hbar$, as we can see, for example for the harmonic oscillator propagator). To get rid of the caustics we can replace the real phase of the WKB method by complex-valued phases. Here we shall report on a related but more explicit approach using elementary properties of coherent states. Coherent states have been used in partial differential equations for a long time, starting with Schrödinger himself and following by many authors (see [14, 6, 3, 20, 25, 19] for applications and historical comments).

We shall follow the presentation given in [25]. Let us now recall some basic definitions concerning coherent states. We start with the ground state of the harmonic oscillator,

$$g_0(x) = \pi^{-d/4} \exp\left(-|x|^2/2\right). \tag{12.21}$$

For $z = (q, p) \in \mathbb{R}^{2d}$, the Weyl–Heisenberg operator of translation by z in phase space is

$$\mathcal{T}_\hbar(z) = \exp\left(\frac{i}{\hbar}(p \cdot x - q \cdot \hbar D_x)\right). \tag{12.22}$$

Let us define the dilation operator $\Lambda_\hbar \psi(x) = \hbar^{-d/4} \psi\left(x\hbar^{-1/2}\right)$. So the coherent state picked on z is defined by $\psi_z = \mathcal{T}_\hbar(z)g_0$. A more explicit expression for ψ_z is

$$\psi_z(x) = e^{\frac{i}{\hbar}\left(p \cdot x - \frac{q \cdot p}{2}\right)}\psi_0(x - q). \tag{12.23}$$

For $\hbar = 1$ we shall denote $\psi_z = g_z$ and $\mathcal{T}_\hbar(z) = \mathcal{T}(z)$.

It is well known that $\{\psi_z\}_{z \in \mathbb{R}^{2d}}$ is an "overcomplete basis" in $L^2(\mathbb{R}^d)$. So for every trace-class observable $\hat{B}$ in $L^2(\mathbb{R}^d)$ we have

$$\mathrm{Tr}(\hat{B}) = (2\pi\hbar)^{-d} \int_Z \langle \hat{B}\psi_z, \psi_z\rangle dz \tag{12.24}$$

where $\langle \cdot, \cdot \rangle$ is the scalar product in $L^2(\mathbb{R}^d)$.

Under rather general assumptions, an asymptotic expansion for the quantum evolution of coherent states, $U(t)\psi_z$, was obtained in [6], with an error term of order $O(\hbar^\infty)$, with some control for large time of order $O(\log(\hbar^{-1}))$ (Ehrenfest time).

Recall that $z_t = (q_t, p_t)$ is the solution of (12.8) starting from $z = (q, p)$. For simplicity it is assumed that $z \in \Omega$, where Ω is as in Section 12.1 and the flow satisfies (12.14). Let us define

$$\delta_t(z) = S(t, z) - \frac{q_t \cdot p_t - q \cdot p}{2} \quad \text{where} \quad S(t, z) = \int_0^t p_s \cdot \dot{q}_s ds - tH_0(z) \tag{12.25}$$

is the classical action. The Jacobi stability matrix, F_t, is the linearized flow associated with (12.8) at the point z_t of the classical trajectory. F_t is also the Hamiltonian flow defined by the quadratic Hamiltonian $K_2(t, \zeta) = \frac{1}{2}\partial^2_{z,z}H(z_t)\zeta \cdot \zeta$ for $\zeta \in \mathbb{R}^{2d}$, where $\partial^2_{z,z}H$ is the Hessian matrix of $H(z)$ in the variables z. We have $F_0 = \mathbb{I}$ and F_t is a symplectic, $2d \times 2d$ matrix. It can be written as four $d \times d$ blocks :

$$F_t = \begin{pmatrix} A_t & B_t \\ C_t & D_t \end{pmatrix}. \tag{12.26}$$

We also need to introduce the quantization of F_t which can be defined as the quantum propagator $\mathcal{M}[F_t]$, *with Planck constant equal to 1*, for the quadratic Hamiltonian $Op_1^w[K_2(t)]$. In particular we have the useful formulas

$$\mathcal{M}[F_t]g_0(x) = \pi^{-d/4}\left(\det(A_t + iB_t)\right)^{-1/2} e^{i\Gamma_t x \cdot x/2}, \tag{12.27}$$

$$\mathcal{M}[F_t]^{-1}Op_1^w[L]\mathcal{M}[F_t] = Op_1^w[L \circ F_t], \tag{12.28}$$

where $\Gamma_t = (C_t + iD_t)(A_t + iB_t)^{-1}$ and L is any classical (smooth) observable defined on Z. Γ_t is a complex symmetric, $d \times d$ matrix, with positive definite imaginary part given as

$$\operatorname{Im}\Gamma_t = (AA' + BB')^{-1} \tag{12.29}$$

where A' denotes the transposed matrix of A.

Let us now state the following result [6, 25].

Theorem 2. *Under the above assumptions, there exists a family of polynomials $\{b_j(t,x)\}_{j\in\mathbb{N}}$ in d real variables $x = (x_1, \cdots, x_d)$, with time-dependent coefficients, such that for every $|t| \leq T$, $\hbar \in {]}0,1]$, $N \in \mathbb{N}$, we have*

$$\left\| U(t)\psi_z - \exp\left(\frac{i\delta_t(z)}{\hbar}\right)\mathcal{T}(z_t)\Lambda_\hbar\mathcal{M}[F_t]\left(\sum_{0 \leq j \leq N} \hbar^{j/2}b_j(t)g_0\right)\right\|_{L^2(\mathbb{R}^d)} \tag{12.30}$$

$$\leq C_N s(T)^{(3N+\epsilon_d)}(1 + T)^{N+1}\hbar^{(N+1)/2}$$

where $s(T)$ can be chosen as in Section 12.2, formula (12.14), and the constant C_N is independent of T and $\hbar$.

Remark 3.
1) This result is proved in [6]. Theorem 2 is applied at finite time to give a proof of the Gutzwiller trace formula.

   In [25] Gevrey type estimates for C_N, N large, have been computed.
2) The polynomials $b_j(t,x)$ can be explicitly computed by induction along the classical trajectories $z(t)$. In particular $b_0(t,x) = \exp\left(-i\int_0^t H_1(z_s)ds\right)$.
3) As for evolution of observables, we get a critical time $T'_\hbar$ for the validity of semiclassical approximation. In the general case $\sigma(T) = e^{\Lambda T}$ and $T'_\hbar = \frac{1-\varepsilon}{6\Lambda}|\log\hbar|$. In the integrable case we can choose $T'_\hbar = \hbar^{\varepsilon - 1/8}$. These validity times are smaller than those found for propagation of observables.

12.4 Semiclassical trace asymptotics

All Hamiltonians considered here satisfy the following general technical assumptions:

(TA1) $\mu_0(z) := 1 + |H_0(z)|$ is a weight function on Z and H is a semiclassical observable with weight μ_0 and principal symbol H_0.

(TA2) E is a fixed energy such that there exists $a < E < b$ and $H_0^{-1}[a,b]$ is a bounded closed set in the phase space Z. Moreover E is noncritical for

H_0, i.e., $\nabla H_0(z) \neq 0$ for every $z \in \Sigma_E$ where $\Sigma_E = H_0^{-1}(E)$. So the Liouville measure is well defined on Σ_E,

$$d\nu_E(z) = \frac{d\Sigma_E(z)}{|\nabla H_0(z)|},$$

where $d\Sigma_E$ is the canonical measure on the hypersurface Σ_E.

Our main goal in this section is to revisit the following spectral distribution:

$$\Xi_{\rho,A}(E,\hbar) = \sum_{j \geq 0} \rho\left(\frac{E_j - E}{\hbar}\right) A_{jj}(\hbar) \tag{12.31}$$

where the Fourier transform $\tilde\rho$ of ρ has a compact support. The ideal ρ should be the Dirac delta function, which needs too much information. So we will try to control large support for $\tilde\rho$. To do that we take $\rho_T(t) = T\rho_1(tT)$ with $T \geq 1$, where ρ_1 is a nonnegative, even, smooth real function, $\int_{\mathbb{R}} \rho_1(t)dt = 1$, $\mathrm{supp}\{\tilde\rho_1\} \subset [-1,1]$, $\tilde\rho_1(t) = 1$ for $|t| \leq 1/2$.

By applying the propagation theorem for coherent states stated in Section 12.3, we can write $\Xi_{\rho_T,A}(E,\hbar)$ as a Fourier integral with an explicit complex phase. The classical dynamics enter the game in a second step, to analyse the critical points of the phase. Let us describe the steps (see [3] for the details).

(i) Modulo a negligible error, we can replace $\hat A$ by $\hat A_\chi = \chi(\hat H)\hat A\chi(\hat H)$ where χ is smooth with support in a small neighborhood of E like $]E - \delta_\hbar, E + \delta_\hbar[$ such that $\lim_{\hbar \to 0} \delta_\hbar = 0$.

(ii) Using an inverse Fourier formula we have the following time-dependent representation:

$$\Xi_{\rho_T,A}(E,\hbar) = \frac{1}{2\pi} \int_{\mathbb{R}} \tilde\rho_1\left(\frac{t}{T}\right) \mathrm{Tr}\left(\hat A_\chi e^{\frac{it}{\hbar}(E - \hat H)}\right) dt. \tag{12.32}$$

(iii) If B is a symbol, then we have $\hat B\psi_z = B(z)\psi_z + \cdots$ where the $\cdots$ are correction terms in half powers of $\hbar$ which depend on the Taylor expansion of B at z.

(iv) Putting all things together, after some computations, we get for every $N \geq 1$:

$$\Xi_{\rho_1,A}(E,\hbar) = (2\pi\hbar)^{-d} \int_{\mathbb{R}_t \times \mathbb{R}_z^{2d}} \tilde\rho_1\left(\frac{t}{T}\right) a^{(N)}(t,z,\hbar) e^{\frac{i}{\hbar}\Psi_E(t,z)} dtdz + \mathcal{R}_{N,T,\hbar}. \tag{12.33}$$

The phase Ψ_E is given by

$$\Psi_E(t,z) = t(E - H_0(z)) + \frac{1}{2}\int_0^t \sigma(z_s - z, \dot z_s)ds$$

$$+ \frac{i}{4}(\mathbb{I}_d - W_t)(\check z - \check z_t) \cdot \overline{(\check z - \check z_t)}, \tag{12.34}$$

with $\check{z} = q + ip$ if $z = (q, p)$ and $W_t = Z_t Y_t^{-1}$ where $Y_t = C_t - B_t + i(A_t + D_t)$, $Z_t = A_t - D_t + i(B_t + C_t)$.

The amplitude $a^{(N)}$ has the property

$$a^{(N)}(t, z, \hbar) = \sum_{0 \leq j \leq N} a_j(t, z)\hbar^j, \tag{12.35}$$

where each $a_j(t, z)$ is smooth, with support in variable z included in the neighborhood $\Omega = H_0^{-1}[E - \delta_E, E + \delta_E]$ of Σ_E, and estimated in t as

$$|a_j(t, z)| \leq C_j s(T)^{6j}(1 + T)^{2j}. \tag{12.36}$$

In particular for $j = 0$ we have

$$a_0(t, z) = \pi^{-d/2}\left[\det(Y_t)\right]^{-1/2} \exp\left(-i \int_0^t H_1(z_s)ds\right). \tag{12.37}$$

The remainder term satisfies

$$\mathcal{R}_{N,T,\hbar} \leq C_N s(T)^{6N+\epsilon_d}(1 + T)^{2N+1}\hbar^{N+1}. \tag{12.38}$$

From the above computations we can easily see that the main contributions as $\hbar \to 0$ in $\Xi_{\rho_T, A}(E, \hbar)$ come from the periods of the classical flow as it is expected. Let us first remark that we have

$$2\mathrm{Im}\Psi_E(t, z) \geq \langle \mathrm{Im}\,\Gamma_t(\Gamma_t + i)^{-1}(\check{z} - \check{z}_t), \Gamma_t + i)^{-1}(\check{z} - \check{z}_t)\rangle.$$

Here $\langle,\rangle$ is the Hermitian product on $\mathbb{C}$. Because of positivity of $\mathrm{Im}\,\Gamma_t$ we get the following lower bound: there exists $c_0 > 0$ such that for every T and $|t| \leq T$ we have

$$\mathrm{Im}\Psi_E(t, z) + |\partial_t\Psi_E(t, z)|^2 \geq c_0\left(|H_0(z) - E|^2 + s(T)^{-4}|z - z_t|^2\right). \tag{12.39}$$

The stationary phase theorem with complex phase [17], vol.1, gives easily the contribution of the 0-period.

Theorem 3. *If T_0 is chosen small enough, such that $T_0 < \sup\{t > 0, \forall z \in \Sigma_E, \Phi^t(z) \neq z\}$, then we have the following asymptotic expansion:*

$$\Xi_{\rho_{T_0}, A}(E, \hbar) \asymp (2\pi\hbar)^{-d} \sum_{j \geq 0} \alpha_{A,j}(E)\hbar^{j+1} \tag{12.40}$$

where the coefficients $\alpha_{A,j}$ do not depend on ρ. In particular

$$\alpha_{A,0}(E) = \int_{\Sigma_E} A(z)d\nu_E(z), \quad \alpha_{A,1}(E) = \int_{\Sigma_E} H_1(z)A(z)d\nu_E(z). \tag{12.41}$$

By using a Tauberian argument [24], a Weyl formula with an error term $O(\hbar^{1-d})$ can be obtained from (12.40).

The contributions of periodic trajectories can be computed if we have some specific assumptions on the classical dynamics. However Petkov–Popov [22] succeed in giving a very general trace formula modulo an error $o(\hbar^{1-d})$ using Hörmander's Fourier integral operator theory. With our coherent states analysis it is possible to recover their result. Let us recall that in [3] this coherent states analysis is used to give a proof of the Gutzwiller trace formula. Let us recall now the rigorous statement.

The main assumption is the following. Let $\mathcal{P}_{E,T}$ be the set of all periodic orbits on Σ_E with periods T_γ, $0 < |T_\gamma| \leq T$ (including repetitions and change of orientation). T_γ^* is the primitive period of γ. Assume that all γ in $\mathcal{P}_{E,T}$ are nondegenerate, i.e., 1 is not an eigenvalue for the corresponding "Poincaré map", P_γ. It is the same to say that 1 is an eigenvalue of F_{T_γ} with algebraic multiplicity 2. In particular, this implies that $\mathcal{P}_{E,T}$ is a finite union of closed path with periods T_{γ_j}, $-T \leq T_{\gamma_1} < \cdots < T_{\gamma_N} \leq T$.

Theorem 4 (Trace Gutzwiller Formula). *Under the above assumptions, for every smooth test function ρ such that* $\mathrm{supp}\{\tilde{\rho}\} \subset] - T, T[$, *the following asymptotic expansion holds true, modulo* $O(\hbar^\infty)$,

$$\Xi_{\rho,A}(E,\hbar) \asymp (2\pi\hbar)^{-d}\tilde{\rho}(0)\sum_{j\geq 0} c_{A,j}(\tilde{\rho})\hbar^{j+1}$$

$$+ \sum_{\gamma \subset \mathcal{P}_{E,T}} (2\pi)^{d/2-1}\exp\left(i\left(\frac{S_\gamma}{\hbar} + \frac{\sigma_\gamma\pi}{2}\right)\right)|\det(\mathbb{I} - P_\gamma)|^{-1/2} \qquad (12.42)$$

$$\left(\sum_{j\geq 0} d_{A,j}^\gamma(\tilde{\rho})\hbar^j\right)$$

where σ_γ is the Maslov index of γ ($\sigma_\gamma \in \mathbb{Z}$), $S_\gamma = \oint_\gamma pdq$ is the classical action along γ, $c_{A,j}(\tilde{\rho})$ are distributions in $\tilde{\rho}$ supported in $\{0\}$, in particular

$$c_{A,0}(\tilde{\rho}) = \tilde{\rho}(0)\alpha_{A,0}(E), \quad c_{A,1}(\tilde{\rho}) = \tilde{\rho}(0)\alpha_{A,1}(E).$$

$d_j^\gamma(\tilde{\rho})$ *are distributions in $\tilde{\rho}$ with support $\{T_\gamma\}$. In particular*

$$d_0^\gamma(\tilde{\rho}) = \tilde{\rho}(T_\gamma)\exp\left(-i\int_0^{T_\gamma^*} H_1(z_u)du\right)\int_0^{T_\gamma^*} A(z_s)ds. \qquad (12.43)$$

Remark 4. By the same method it is possible as well to consider integrable systems to get the Berry–Tabor formula or more generally systems satisfying the clean intersection property [3].

For larger time we can use the time-dependent estimates given above to improve the remainder estimate in the Weyl asymptotic formula, as Volovoy did for elliptic operators on compact manifolds [27]. For that, let us introduce some control on the measure of the set of periodic path. We call this property condition (NPC). Let $J_E =]E - \delta, E + \delta[$ be a small neighborhood of energy E and $s_E(T)$ an increasing function as in (12.14) for the open set $\Omega_E = H_0^{-1}(J_E)$. We assume for simplicity here that s_E is either an exponential $(s_E(T) = \exp(\Lambda T^b), \Lambda > 0, b > 0)$ or a polynomial $(s_E(T) = (1 + T)^a, a \geq 1)$.

The condition is the following:

(NPC) $\forall T_0 > 0$, there exist positive constants $c_1, c_2, \kappa_1, \kappa_2$ such that for all $\lambda \in J_E$ we have

$$\nu_\lambda \left\{ z \in \Sigma_\lambda, \ \exists t, \ T_0 \leq |t| \leq T, \ |\Phi^t(z) - z| \leq c_1 s(T)^{-\kappa_1} \right\} \leq c_2 s(T)^{-\kappa_2}.$$

$$(12.44)$$

The following result, proved by using stationary phase arguments, estimates the contribution of the "almost periodic points":

Proposition 1. *For all* $0 < T_0 < T$, *let us denote* $\rho_{T_0 T}(t) = (1 - \rho_{T_0})(t)\rho_T(t)$, *where* $0 < T_0 < T$. *Then we have*

$$\Xi_{\rho_{T_0 T}, A}(E, \hbar) \leq C_3 s(T)^{-\kappa_3} \hbar^{1-d} + C_4 s(T)^{\kappa_4} \hbar^{2-d} \qquad (12.45)$$

for some positive constants $C_3, C_4, \kappa_3, \kappa_4$.

Let us now introduce the integrated spectral density

$$\sigma_{A,I}(\hbar) = \sum_{E_j \in I} A_{jj}(\hbar) \qquad (12.46)$$

where $I = [E', E]$ is such that for some $\lambda' < E' < E < \lambda$, $H_0^{-1}[\lambda', \lambda]$ is a bounded closed set in Z and E', E are regular for H_0. We have the following two-term Weyl asymptotics with a remainder estimate.

Theorem 5. *Assume that there exist open intervals* J_E *and* $J_{E'}$ *satisfying the condition (NPC). Then we have*

$$\sigma_{A,I}(\hbar) = (2\pi\hbar)^{-d} \int_{H_0^{-1}(I)} A(z)dz - (2\pi)^{-d}\hbar^{1-d}$$
$$\times \left(\int_{\Sigma_E} A(z)H_1(z)d\nu(z) - \int_{\Sigma_{E'}} A(z)H_1(z)d\nu(z) \right) + O\left(\hbar^{1-d}\eta(\hbar)\right) \qquad (12.47)$$

where $\eta(\hbar) = |\log(\hbar|^{-1/b}$ *if* $s_E(T) = \exp(\Lambda T^b)$ *and* $\eta(\hbar) = \hbar^\varepsilon$, *for some* $\varepsilon > 0$, *if* $s_E(T) = (1 + T)^a$. *Furthermore if* $I_{E, \delta_1, \delta_2}(\hbar) = [E + \delta_1\hbar, E + \delta_2\hbar]$ *with* $\delta_1 < \delta_2$, *then we have*

$$\sum_{E + \delta_1\hbar \leq E_j \leq E + \delta_2\hbar} A_{jj}(\hbar) = (2\pi\hbar)^{1-d}(\delta_2 - \delta_1) \int_{\Sigma_E} A(z)d\nu_E + O\left(\hbar^{1-d}\eta(\hbar)\right).$$

$$(12.48)$$

Remark 5. (i) Formula 12.48 was first proved in [2], without remainder estimate.

(ii) Theorem 5 is deduced from Proposition 1 by a Tauberian argument as in [23, 27, 2]

As was done in [5], by the same technique used above it is possible to give a semiclassical asymptotic expansion for the Wilkinson variance introduced before. Let us first write the time-dependent representation formula

$$
\mathcal{V}_{(\rho,g,A)}(\hbar, E, \tau) = \frac{\hbar^{-1}}{4\pi^2} \iint_{\mathbb{R}\times\mathbb{R}} \tilde{\rho}(t)\tilde{g}(u - t/2)
$$
$$
\times \mathrm{Tr}\left(\hat{A}_u \hat{A} \exp -\frac{it}{\hbar}\left(\hat{H} - E + \frac{\hbar\tau}{2}\right)\right) e^{iu\tau}\, dt\, du\,.
\tag{12.49}
$$

We know from Section 12.2 that $\hat{A}_u := U(-u)\hat{A}U(u)$ is a semiclassical observable, with uniform estimates for $|\tau| \leq C\log(\hbar^{-1})$. So for ρ as in (12.42) (smooth with compact support) we can choose g Gaussian. It is enough to assume that g is a smooth function such that $|\tilde{g}(\tau)| \leq Ce^{-|\tau|/\varepsilon}$, for $\varepsilon > 0$ small enough (depending only on $s(T)$). Then under the same conditions as for the Gutzwiller trace formula we have

Theorem 6 (Wilkinson Trace Formula).

$$
\mathcal{V}_{(\rho,g,A)}(\hbar, E, \tau) \asymp (2\pi\hbar)^{-d} \sum_{j\geq 0} c_{A,j}(\tau, \tilde{\rho}, g)\hbar^j + \sum_{\gamma\in\mathcal{P}_{E,T}} (2\pi)^{d/2-1}
$$
$$
\times \exp\left(i\left(\frac{S_\gamma}{\hbar} + \frac{\sigma_\gamma\pi}{2}\right)\right)|\det(\mathbb{I} - P_\gamma)|^{-1/2}\left(\sum_{j\geq 0} d^\gamma_{A,j}(\tau, \tilde{\rho}, g)\hbar^{j-1}\right)
\tag{12.50}
$$

where $c_{A,j}(\tau, \tilde{\rho}, g)$ are distributions in $\tilde{\rho}$ supported in $\{0\}$, in particular

$$
c_{A,0}(\tau, \tilde{\rho}, g) = \tilde{\rho}(0) \int \tilde{g}(t)e^{it\tau} C_A(E,t)dt
\tag{12.51}
$$

where $C_A(E,t)$ is the classical autocorrelation function

$$
C_A(E,t) = \int_{\Sigma_E} A(z)A(\Phi^t(z))d\nu_E(z)\,.
\tag{12.52}
$$

Moreover, $d^\gamma_j(\tau, \tilde{\rho}, g)$ are distributions in $\tilde{\rho}$ with support $\{T_\gamma\}$. In particular

$$
d^\gamma_0(\tau, \tilde{\rho}, g) = \tilde{\rho}(T_\gamma)\exp\left(-i\int_0^{T^*_\gamma}(H_1(z_u) + \tau/2)du\right)
$$
$$
\cdot \int_{\mathbb{R}} C^\gamma_A(u)\tilde{g}\left(\frac{T_\gamma}{2} - u\right)e^{iu\tau}\, du\,.
\tag{12.53}
$$

where $C^\gamma_A(u) = \int_0^{T^*_\gamma} A(z_{u+t})A(z_t)dt$ is the autocorrelation function along γ.

Remark 6. In [7] a similar result was found in the theory of linear response. As a distribution in g, $d_0^\gamma(\tilde\rho, g)$ can be conveniently written with the Fourier coefficient of the T_γ^*-periodic function $C_A^\gamma(u)$. So we get

$$\int_\mathbb{R} C_A^\gamma(u)\tilde g\Big(u - \frac{T_\gamma}{2}\Big)e^{iu\tau}du$$
$$= \frac{2\pi}{T_\gamma^*}\sum_{k\in\mathbb{Z}} C_{A,k}^\gamma g\Big(\frac{2\pi k}{T_\gamma^*} + \tau\Big)\exp\Big(i\Big(\frac{2\pi k}{T_\gamma^*} + \tau\Big)\frac{T_\gamma}{2}\Big) \tag{12.54}$$

with the Fourier decomposition $C_A^\gamma(u) = \dfrac{1}{T_\gamma^*}\sum_{k\in\mathbb{Z}} C_{A,k}^\gamma \exp\Big(\dfrac{2i\pi ku}{T_\gamma^*}\Big)$.

This shows, in particular, that $d_0^\gamma(\tau, \tilde\rho, g)$ is a distribution in g supported in the discrete set $\{(2\pi k/T_\gamma^*) + \tau, \ k\in\mathbb{Z}\}$.

12.5 Quantum ergodicity and mixing

This section revisits some results first proved by Sunada [26] and Zelditch [29] for compact manifolds. Our presentation is somehow different and some estimates are improved. Let us begin with a rough estimate concerning the matrix elements $A_{jk}(\hbar)$ and $\omega_{j,k}(\hbar)$. Let $I = [E', E]$ be an energy interval, an observable $A \in \mathcal{O}(\mu)$ for some weight μ. Assume that for some $\lambda' < E' < E < \lambda$, $H_0^{-1}[\lambda', \lambda]$ is compact. Let us choose χ, a smooth cutoff supported in $]\lambda', \lambda[$, such that $\chi = 1$ on I. Let us introduce the new observable $\hat A_\chi = \chi(\hat H)\hat A\chi(\hat H)$. Then starting from the equality

$$\frac{i}{\hbar}\langle[\hat A_\chi, \hat H]\varphi_j, \varphi_k\rangle = \omega_{j,k}(\hbar)A_{jk}(\hbar),$$

by induction, we get for every $N \geq 1$, $|\omega_{j,k}(\hbar)|^N|A_{jk}(\hbar)| \leq C_N$ where C_N is independent of $\hbar$.

The following result, proved in [5] and following Helton's trick, show that a single nonperiodical trajectory disturbs very much the energy spectrum.

Theorem 7. *Suppose that on Σ_E there exists at least one periodical classical trajectory. Let $0 < \hbar_n$ such that $\lim_{n\to 0} \hbar_n = 0$. Then for every $\varepsilon > 0$ and $c > 0$, the set*

$$\big\{\omega_{j,k}(\hbar_n),\ n \in \mathbb{N},\ E_j, E_k \in [E - c\hbar_n^{1-\varepsilon}, E + c\hbar_n^{1-\varepsilon}]\big\}$$

is dense in $\mathbb{R}$.

Sketch of proof: For every observable $A, B \in \mathcal{O}(0)$ and every integrable function f, let us introduce $\hat A_f = \int_\mathbb{R} f(t)\hat A_t dt$ where $\hat A_t = U(-t)\hat A U(t)$. An easy formal computation, assuming for simplicity that φ_j is an orthonormal basis in $L^2(\mathbb{R}^d)$, gives

$$\mathrm{Tr}(\hat{A}_f \hat{B}) = \sum_{j,k} A_{jk}(\hbar) B_{kj}(\hbar) \tilde{f}(\omega_{j,k}(\hbar)) . \tag{12.55}$$

Assume that $\tilde{f}(\omega_{j,k}(\hbar_n)) = 0$ for all j, k, n. Taking the limit $n \to +\infty$, we get $\int_Z A(\Phi^t(z)B(z)f(t)dtdz = 0$. Using existence of one nonperiodical trajectory we construct suitable observables such that for every smooth function $k(t)$, with compact support, we can get $\int_{\mathbb{R}} k(t)f(t)dt = 0$, hence $f \equiv 0$.   □

We shall now consider some specific dynamical properties for the flow on the energy shell Σ_E, equipped with the flow invariant Liouville measure. For $A \in L^1(\Sigma_E)$, let us denote its average $\langle A \rangle_E = \int_{\Sigma_E} A(z)d\bar{\nu}_E(z)$.

Definition 2.

(i) Φ^t is ergodic on Σ_E if for every measurable function A on Σ_E we have

$$\{\forall t \in \mathbb{R}, A \circ \Phi^t = A\} \Longleftrightarrow \{A = \text{constant a.e on } \Sigma_E\} . \tag{12.56}$$

(ii) Φ^t is weakly mixing on Σ_E if for every $A \in L^2(\Sigma_E)$ we have

$$\lim_{T \to \infty} \frac{1}{2T} \int_{-T}^{T} C_A(E,t)dt = \langle A \rangle_E^2 . \tag{12.57}$$

Let us introduce the unitary group in $L^2(\Sigma_E)$ defined as $(\mathcal{U}(t)A)(z) = A(\Phi^t(z))$. A consequence of spectral theory is existence, for every $A \in L^2(\Sigma_E)$, of a finite Borel measure μ_A on the real axis $\mathbb{R}$ such that

$$\langle \mathcal{U}(t)A.A \rangle_E = \int_{\mathbb{R}} e^{it\theta} d\mu_A(\theta).$$

Then we have the following spectral characterizations:

$$\{\Phi^t \text{ ergodic}\} \Longleftrightarrow \{\text{supp}\{\mu_A\} = \{0\} \Rightarrow A = \text{constant}\} \tag{12.58}$$

$$\{\Phi^t \text{ weakly mixing}\} \Longleftrightarrow \{A \in L^2(\Sigma_E), \langle A \rangle_E = 0$$
$$\Longrightarrow \mu_A \text{ is continuous}\} . \tag{12.59}$$

Let us consider a small slice of energy, $I_{\hbar}^{\varepsilon} = [E - c\hbar^{1-\varepsilon}, E - c\hbar^{1-\varepsilon}]$, where $\varepsilon \geq 0$. The quantum analogue of the Liouville measure and the spectral measures for a flow can be defined as follows.

We have $\mathrm{Spec}[\hat{H}] \cap I_{\hbar}^{\varepsilon} = \{E_1 \leq E_2 \leq \cdots \leq E_{N_{\hbar}^{\varepsilon}}\}$ with multiplicities and we introduce the spectral projectors $\Pi_{\hbar}^{\varepsilon}$ of $\hat{H}$ on $I_{\hbar}^{\varepsilon}$ so that $\mathrm{Tr}(\Pi_{\hbar}^{\varepsilon}) = N_{\hbar}^{\varepsilon}$. Using Weyl asymptotics, it is not difficult to prove, for every $\varepsilon > 0$,

$$\lim_{\hbar \to 0} \frac{1}{N_{\hbar}^{\varepsilon}} \mathrm{Tr}(\Pi_{\hbar}^{\varepsilon} \hat{A}) = \int_{\Sigma_E} A(z)d\bar{\nu}_E(z) . \tag{12.60}$$

It is well known that we can modify the quantization $\hat{A}$ with an error $O(\hbar)$ such that $A \mapsto A_{jj}$ is a positive Radon measure. This is done by taking $A_{jj} = \langle Op_\hbar^{aw} A \varphi_j, \varphi_j \rangle$, where $Op_\hbar^{aw}$ is the anti-Wick quantization, easily defined using coherent states analysis of Section 12.1: $Op_\hbar^{aw} \eta(q) = (2\pi\hbar)^{-d} \int_Z A(z) \langle \eta, \psi_z \rangle \psi_z(q) dz$. So we can write $A_{jj}(\hbar) = \int_Z A d\upsilon_j(z)$ where $d\upsilon_j$ is a probability measure on Z. Then, for every $\varepsilon > 0$, we have for the weak convergence of measures

$$\lim_{\hbar \to 0} \frac{1}{N_\hbar^\varepsilon} \sum_{\{E_j \in I_\hbar^\varepsilon\}} d\upsilon_j = d\bar{\nu}_E \,. \tag{12.61}$$

We can also define, modulo $O(\hbar^\infty)$, the Borel measures $dm_{A,\hbar}$,

$$\int_\mathbb{R} f(\theta) dm_{A,\hbar}(\theta) = \frac{1}{N_\hbar^\varepsilon} \sum_{E_j, E_k \in I_\hbar^\varepsilon} f(\omega_{j,k}(\hbar)) |A_{jk}(\hbar)|^2 \,. \tag{12.62}$$

We also have

$$\int_\mathbb{R} f(\theta) dm_{A,\hbar}(\theta) = \frac{1}{N_\hbar^\varepsilon} \mathrm{Tr}(\Pi_\hbar^\varepsilon \hat{A}_f \Pi_\hbar^\varepsilon \hat{A}) \,. \tag{12.63}$$

Taking the classical limit we have the weak convergence of $dm_{A,\hbar}$ to $d\mu_A$,

$$\lim_{\hbar \to 0} \int_\mathbb{R} f(\theta) dm_{A,\hbar}(\theta) = \int_\mathbb{R} f(\theta) d\mu_A(\theta) \,. \tag{12.64}$$

The ergodic quantum theorem can be stated as follows:

Theorem 8. *Let us assume here $\varepsilon = 0$ and denote $N_\hbar^0 = N_\hbar$, $I_\hbar^0 = I_\hbar^0$. If the classical system is ergodic on Σ_E, then we have*

$$\lim_{\hbar \to 0} \frac{1}{N_\hbar} \sum_{\{E_j \in I_\hbar\}} \left| \int A d\upsilon_j - \langle A \rangle_E \right|^2 = 0 \,. \tag{12.65}$$

Furthermore, we have some estimate on the rate of classical ergodicity, and if the (NPC) condition is fulfilled, *then we get a control of the quantum ergodicity. So let us assume that for some $0 < a \leq 1$ we have*

$$\int_{\Sigma_E} \left| \frac{1}{2T} \int_{-T}^T A(\Phi^t(z)) dt - \langle A \rangle_E \right|^2 d\nu_E(z) = O(T^{-a}) \,; \tag{12.66}$$

then there exists $C > 0$ such that

$$\frac{1}{N_\hbar} \sum_{\{E_j \in I_\hbar\}} \left| \int A d\upsilon_j - \langle A \rangle_E \right|^2 \leq C |\log(\hbar)|^{-a} \,. \tag{12.67}$$

The inverse problem was discussed by Sunada [26] and Zelditch [29]. It is still open. But a partial result can be proved by considering contribution of nearby nondiagonal matrix elements.

Theorem 9. *Let us assume that the condition* (NPC) *is satisfied. Then the classical flow is ergodic on Σ_E if and only if for every $A \in \mathcal{O}(0)$ such that $\langle A \rangle_E = 0$ and every $\alpha :]0,1] \mapsto]0,+\infty[$ such that $\lim_{\hbar \to 0} \alpha(\hbar) = 0$, $\lim_{\hbar \to 0} \alpha(\hbar)|\log(\hbar)| = +\infty$, we have*

$$\lim_{\hbar \to 0} \frac{1}{N_\hbar} \sum_{\substack{\{E_j \in I_\hbar, \\ |\omega_{jk}(\hbar)| \leq \alpha(\hbar)\}}} |A_{jk}(\hbar)|^2 = 0 \,. \tag{12.68}$$

We can also get a similar result for weak-mixing systems.

Theorem 10. *Let us assume that the condition* (NPC) *is satisfied. Then the classical flow is weakly-mixing on Σ_E if and only if for every $\lambda \in \mathbb{R}$, every $A \in \mathcal{O}(0)$ such that $\langle A \rangle_E = 0$ and every $\alpha :]0,1] \mapsto]0,+\infty[$ such that $\lim_{\hbar \to 0} \alpha(\hbar) = 0$, $\lim_{\hbar \to 0} \alpha(\hbar)|\log(\hbar)| = +\infty$, we have*

$$\lim_{\hbar \to 0} \frac{1}{N_\hbar} \sum_{\substack{\{E_j \in I_\hbar, \\ |\omega_{jk}(\hbar)-\lambda| \leq \alpha(\hbar)\}}} |A_{jk}(\hbar)|^2 = 0 \,. \tag{12.69}$$

The starting points to prove these results are the following trace formulae. For the ergodic case we use

$$\sum_{\{E_j, E_k \in I_\hbar\}} \left| A_{jk}(\hbar) \widetilde{M_T}(\omega_{jk}(\hbar)) \right|^2 = \mathrm{Tr}\left(\hat{A}_{M_T} \right)^2 \tag{12.70}$$

with $M_T(u) = \frac{1}{2T} \mathbb{I}_{[-T,T]}$ and for the weak-mixing case,

$$\sum_{\{E_j, E_k \in I_\hbar\}} |A_{jk}(\hbar)|^2 \widetilde{M_T^{(2)}}(\omega_{jk}(\hbar) - \lambda)) = \mathrm{Tr}\left(\int_{\mathbb{R}} e^{it\lambda} \hat{A}_t \hat{A}_{M_T^{(2)}}(t) dt \right) \tag{12.71}$$

with $M_T^{(2)} = M_T \star M_T$ ($\star$ denotes the convolution product). Then the two theorems are proved by computing the semiclassical limits of the right-hand side carefully. $\square$

Remark 7. The definition of quantum ergodicity or quantum weak mixing proposed by Sunada and Zelditch is property (12.65) or (12.69) with $a(\hbar) \equiv 0$. It is not known if these definitions are equivalent to the classical ones.

Acknowledgment. The author thanks M. Combescure for her comments on this paper.

References

1. Bouzouina A., Robert D., Uniform semiclassical estimates for the propagation of observables, *Duke Mathematical Journal* **111**:2 (2002), 223–252.

2. Brummelhuis R., Uribe A., A semiclassical trace formula for Schrödinger operators, *Comm.Math. Phys.* **136** (1991), 567–584.

3. Combescure M., Robert D., Ralston J., A proof of the Gutzwiller Semiclassical Formula Using Coherent States Decomposition, *Commun. Math. Phys.* **202** (1999), 463–480.

4. Combescure M., Robert D., Distribution of matrix elements and level spacings for classically chaotic systems, *Ann. Inst. Henri Poincaré* **61**:4 (1994), 443–483.

5. Combescure M., Robert D., Semiclassical sum rules and generalized coherent states, *J. M. P.* **36**:12 (December 1995).

6. Combescure M., Robert D., Semiclassical spreading of quantum wave packet and applications near unstable fixed points of the classical flow, *Asymptotic Analysis* **14** (1997), 377–404.

7. Combescure M., Robert D., Rigorous semiclassical results for the magnetic response of an electron gas, *Reviews in Math. Phys.* **13**:9 (2001), 1055–1073.

8. Combescure M., Robert D., Semiclassical results in the linear response theory, *Annals of Physics* **305** (2003), 45–59.

9. De Bièvre S., Robert D., Semiclassical propagation and the $\log \hbar^{-1}$ time-barier, *I. M. R. N.* **12** (2003), 667–696.

10. Dozias S., thesis, DMI. ENS., Paris, 1994.

11. Duistermaat J.J., Guillemin V., The spectrum of positive elliptic operators and periodic bicharacteristics, *Inv. Math.* **29** (1975), 39–79.

12. Folland G.B., *Harmonic Analysis in Phase Space*, Annals of Math. Studies, Vol. 122, Princeton University Press, 1989.

13. Gutzwiller M., Periodic orbits and classical quantization conditions, *J. Math. Phys.* **12**:3 (1971), 343–358.

14. Hagedorn G., Semiclassical quantum mechanics, *Ann. Phys.* **135** (1981), 58–70; *Ann. Inst. Henri Poincaré* **42** (1985), 363–374.

15. Helffer B., Martinez A., Robert D., Ergodicité et limite semi-classique, *CMP* **109** (1987), 313–326.

16. Hörmander L, The spectral function of an elliptic operator, *Acta. Math.* **121** (1968), 193–218.

17. Hörmander L. *The Analysis of Linear Partial Differential Operators*, Vols. I–IV, Springer-Verlag, 1983–85.

18. Ivrii V., *Microlocal Analysis and Precise Spectral Asymptotics*, Springer, 1998.

19. Littlejohn R.G., The semiclassical evolution of wave packets, *Physics Report* **138**:4,5 (1986), 193–291.

20. Paul T., Semi-classical methods with emphasis on coherent states, in *IMA Volumes in Mathematics and Applications*, J. Rauch and B. Simon, eds., Vol. 95, Springer, 1997, 51–97.

21. Paul T., Uribe A., Sur la formule semi-classique des traces, *Note CRAS* **313**:I (1991), 217–222.

22. Petkov V., Popov G., Semiclassical trace formula and clustering of eigenvalues for Schrödinger operators, *Ann. I.H.P., Sect. Phys. Th* **68** (1998), 17–83.

23. Petkov V., Robert D., Asymptotique semi-classique du spectre d'hamiltoniens quantiques et trajectoires classiques périodiques, *Comm. in PDE* **10**:4 (1985), 365–390.

24. Robert D., *Autour de l'Approximation Semi-classique*, Birkhäuser, Prog. Math., Vol. 68., 1987.

25. Robert D., *Remarks on asymptotic solutions for time dependent Schrödinger equations.* in Optimal control and partial differential equations, IOS Press, 188–197, (2001).
26. Sunada T., Quantum ergodicity, in *Progress in Inverse Spectral Geometry*, Andersson, Stig I. et al., eds., Birkhäuser, Basel Trends in Mathematics (1997), 175–196.
27. Volovoy A., Improved two-terms asymptotics for the eigenvalue distribution of an elliptic operator on a compact manifold, *Comm. in P.D.E.* **15**:11 (1990), 1509–1563.
28. Wilkinson M., A semi-classical sum rule for matrix elements of classically chaotic systems, *J. Phys. A: Math. Gen.* **20** (1987), 2415–2423.
29. Zelditch S., Quantum mixing, *JFA* **140** (1996), 68–86; and *Quantum Dynamics from the Semi-classical Point of View*, unpublished lectures, Institut Borel, IHP, Paris, 1996.

13

Nonlinear Time-dependent Schrödinger Equations with Double-Well Potential

A. Sacchetti

Dipartimento di Matematica
Università di Modena e Reggio Emilia
Via Campi 213/B
I–41100 Modena, Italy

Summary. We consider a class of Schrödinger equations with a symmetric double-well potential and a nonlinear perturbation. We show that, under certain conditions, the reduction of the time-dependent equation to a two-mode equation gives the dominant term of the solution with a precise estimate of the error.

13.1 Introduction

Recently the nonlinear time-dependent Schrödinger equation (hereafter NLS)

$$\begin{cases} i\hbar\frac{\partial\psi}{\partial t} = \left[-\hbar^2\Delta + V\right]\psi + \epsilon f\psi, \ \epsilon \in \mathrm{R} \\ \psi(t,x)|_{t=0} = \psi^0(x) \end{cases} \quad x \in \mathrm{R}^n, \ n \geq 1, \quad (13.1)$$

where $V = V(x)$ is a given potential and $\epsilon f(|\psi|, x)$ is a nonlinear perturbation with strength ϵ, has received an increasing interest. Indeed, applications of such an equation can be found in many fields: laser physics, optical fibers, dynamics of chemical reactions and bosonic matters, to name some of them. More precisely, the nonlinear Schrödinger equation arises in various physical contexts concerning nonlinear waves, such as propagation of a laser beam in a medium whose index of reflection is sensitive to the wave amplitude, water waves at the free surface and plasma waves [13]. NLS also appears in the description of the Bose–Einstein condensates, a context where (13.1) is often called the Gross–Pitaevskii equation [1], and in the description of the localization effect in a pyramidal molecule like ammonia NH_3 [8], [15].

The last few years have witnessed a rapid development in research on NLS-related applications, and this has created an enormous number of new mathematical problems for mathematicians. Despite such a large interest the theoretical treatment of these equations is very far from complete, in fact it practically begins with the seminal work by Ginibre and Velo [4], and, when the potential V is not zero, very few rigorous results for the study of the solution of the time-dependent equation (13.1) exist. In fact, these results concern the existence of solutions asymptotically given by solitary wave functions when the discrete spectrum of the linear Schrödinger operator has only one nondegenerate eigenvalue [14], [17].

In this paper I summarize some recent results [6], [11], [12] where we consider the NLS equation of the form (13.1) with a *symmetric double-well potential*, that is,

$$V(x', -x_n) = V(x', x_n), \quad x = (x', x_n), \ x' \in \mathrm{R}^{n-1}, \ x_n \in \mathrm{R}. \quad (13.2)$$

In particular, we prove that the *beating motion*, usually observed in the linear Schrödinger equation with a double-well potential, disappears when the strength of the nonlinear term is larger than a critical value. In order to be more precise let us briefly recall the basic notion of beating motion for a double-well Schrödinger equation. Let $E_+ < E_-$ be the two lowest eigenvalues of H, where H is the linear operator formally defined on $L^2(\mathrm{R}^n, dx)$ as

$$H = -\hbar^2 \Delta + V(x), \quad \Delta = \sum_{j=1}^{n} \frac{\partial^2}{\partial x_j^2},$$

with associated normalized eigenvectors φ_+ and φ_-. By means of a suitable choice of gauge we can always choose the eigenvectors such that they are real-valued and

$$\varphi_\pm(x', \pm x_n) = \pm\varphi_\pm(x', x_n).$$

If the state ψ is initially prepared on the first two states, i.e.,

$$\Pi_c \psi^0 = 0$$

where Π_c denotes the projector operator defined as

$$\Pi_c = 1 - \Pi_- - \Pi_+, \quad \Pi_\pm = \langle \varphi_\pm, \cdot \rangle \varphi_\pm, \quad (13.3)$$

then equation (13.1) has a solution given by

$$\psi(t, x) = c_+ e^{-iE_+ t/\hbar} \varphi_+ + c_- e^{-iE_- t/\hbar} \varphi_-$$

and the *center of mass* of the wavefunction ψ, defined as

$$\langle g \rangle^t = \langle \psi(t, \cdot), g(\cdot)\psi(t, \cdot) \rangle,$$

is a periodic function with period

$$T = \frac{\pi \hbar}{\omega}, \quad \omega = \frac{E_- - E_+}{2}.$$

Here, $g \in L^2(\mathrm{R}^n, dx)$ is any bounded function such that $g(x', -x_n) = -g(x', x_n)$, that locally it behaves like x_n. In particular, the *center of mass* periodically assumes positive and negative values, that is we observe the so-called *beating motion* of the wavefunction.

The question we discuss in this work is the following one: *is such a beating motion still present when we restore the nonlinear term?* This problem, as we'll see later, plays a basic role in understanding the localization effect for some relevant physical systems, e.g., existence of chiral configurations in symmetrical molecules like ammonia.

We emphasize that the unit time, for our purposes, is the beating period T. Hence, we will introduce a *slow time* defined as

$$\tau = \frac{t\omega}{\hbar}. \tag{13.4}$$

In this paper we will show that, under certain circumstances, when the nonlinear parameter ϵ is larger than a critical value, then the beating motion disappears and the wavefunction ψ remains essentially localized within one of the two wells. This result follows from the asymptotic behavior of the solution of equation (13.1) in the semiclassical limit, with a precise estimate of the error uniformly in the interval $\tau \in [0, \tau']$ for any fixed τ'.

13.2 Description of the model

Here, we essentially study the NLS equation (13.1) having in mind two applications: the description of the inversion motion of pyramidal molecules and the dynamics of Bose–Einstein condensate states trapped in a double-well potential.

13.2.1 Inversion motion of pyramidal molecules

The existence of a well-defined molecular structure for a symmetric molecule is an old and ongoing problem in chemistry [15]. It was clear from the beginning that the action of the environment would be the basic cause of this effect, since from a quantum mechanics point of view an isolated symmetric molecule has no structure.

As a particularly interesting example, we have pyramidal molecules such as ammonia, NH_3. Such molecules should have a pyramidal structure with the nuclei H in a triangular basis and the nucleus N in a vertex; the chirality will depend on the choice of the vertex. Quantum mechanics predicts symmetrical molecules with the nucleus N delocalized in both vertices. Indeed, in the Bohr-Oppenheimer approximation the nucleus N is affected by a double-well potential, and we have even and odd stationary states and beating states. In fact, physically we always observe the structure of a pyramidal molecule for pressure large enough.

Davies [2] suggested a nonlinear Stark-type model in order to consider the interaction of a single molecule with the other molecules of a given gas. In the mean field approximation this term could be described by means of the effective potential depending on the wavefunction itself [5], [8], [9]:

$$f = f(|\psi|) = \epsilon\langle g\rangle g, \quad \langle g\rangle = \langle\psi, g\psi\rangle, \quad \epsilon \in \mathrm{R}, \tag{13.5}$$

where g is a bounded odd function, $g(x', -x_n) = -g(x', x_n)$, $\langle g\rangle$ measures the dipole of the molecule and ϵ measures the strength of the dipole-dipole interaction. Hence, the inversion motion of the atom N is described by means of an NLS equation of the form

$$i\hbar\frac{\partial\psi}{\partial t} = \left[-\hbar^2\Delta + V(x) + \epsilon\langle\psi, g\psi\rangle g(x)\right]\psi. \tag{13.6}$$

Here, we prove that the beating motion of the *center of mass*, defined by $\langle q\rangle$, disappears when the nonlinear strength is larger than a critical value, giving so the chiral configuration of the ammonia molecule.

13.2.2 Bose–Einstein condensate states in a double-well trap

Bose–Einstein condensation (BEC), predicted more than 70 years ago, obeys an NLS equation, known in the literature as the Gross–Pitaevskii equation [1], of the form

$$i\hbar\frac{\partial\psi}{\partial t} = \left[-\hbar^2\Delta + V(x) + \epsilon|\psi|^2\right]\psi. \tag{13.7}$$

In the case of two Bose–Einstein condensates, under the effect of a double-well trap with a barrier between the two condensates, an interesting application is the possible occurrence of Josephson-type effects [10]. Usually, the description of the dynamics for a Bose–Einstein condensate in a double-well trap is reduced to a nonlinear two-mode equation for the time-dependent amplitudes, which admits explicit solution. Here, we put on a full rigorous basis the results obtained by [10] in the two-level approximation giving the proof of the stability of such a reduction.

13.3 Assumptions and parameters

We assume that the double-well potential $V(x)$ is a function regular enough satisfying the condition (13.2) and with two (nondegenerate) minima at $x_\pm = (0, \pm d)$, $d > 0$, such that $V(x) > V(x_\pm)$, $\forall x \neq x_\pm$; moreover, we assume also that $V(x)$ has the following behavior at infinity:

$$\liminf_{|x|\to\infty} V(x) > V(x_\pm).$$

Hence, the discrete spectrum of the linear Schrödinger operator H is not empty if $\hbar$ is small enough.

In this paper we consider the semiclassical limit of the NLS equation (13.1) where the actual semiclassical parameter is the splitting $\omega \ll 1$ between the two wells. More precisely, we consider the simultaneous limit of small splitting

$$\omega \to 0 \tag{13.8}$$

and small nonlinear perturbation strength

$$\epsilon \to 0$$

such that the effective nonlinearity parameter defined as

$$\eta = \frac{c\epsilon}{\omega} = O(1) \tag{13.9}$$

goes to a constant, where c will be defined later in equation (13.24).

The condition (13.8) could be obtained in different ways: we take the limit of $\hbar$ that goes to zero and fixed double-well potential [6], [12], or we fix $\hbar = 1$ and we consider the limit of a large barrier between the two wells [11]. In fact, it is well known that the splitting ω satisfies the asymptotic behavior $\omega = O(e^{-\rho/\hbar})$ where ρ is the Agmon distance between the two wells.

13.4 Main result

Our main result is the following one.

Proposition. *Let ψ^0 be such that*

$$\Pi_c \psi^0 = 0 \tag{13.10}$$

where Π_c is the projection operator defined in (13.3). In the semiclassical limit (13.9) the destruction of the beating motion follows for any value of the effective nonlinearity parameter η larger than a critical value.

Remark. From this fact and since the nonlinear strength is directly related to the pressure of the ammonia gas [8], it follows that this model is able to describe the appearance of chiral configuration for the ammonia molecule when the gas pressure is larger than a critical value.

Remark. In the semiclassical limit of small $\hbar$ and fixed double-well potential the above proposition holds for any dimension $d \geq 1$ if the nonlinear perturbation is given by (13.5) [6]. In contrast, in the case of a BEC-type NLS equation (13.7) the above proposition holds for dimension $d = 1$ [12].

Remark. In the case of fixed $\hbar$ and in the limit of a large barrier between the two wells we have to introduce some further assumptions [11]; more precisely, we assume that the discrete spectrum of the linear operator consists of only two nondegenerate eigenvalues and that the dissipative estimate of the type

$$\left\| e^{-iHt} \Pi_c \varphi \right\|_p \le C[1 + |t|]^{-d\left(\frac{1}{2} - \frac{1}{p}\right)} \|\varphi\|_{p'}. \tag{13.11}$$

for any p and p', $\frac{1}{p} + \frac{1}{p'} = 1$ and $2 \le p \le \infty$, and where C is a positive constant, uniformly holds. For such a reason we have to assume that the dimension should be $d = 1$ or $d = 3$.

13.5 Idea of the proof

Here, for the sake of definiteness, we restrict ourselves to the one-dimensional semiclassical limit of the BEC-type NLS equation

$$\begin{cases} i\hbar \frac{\partial \psi}{\partial t} = -\hbar^2 \frac{\partial^2 \psi}{\partial x^2} + V(x)\psi(x) + \epsilon |\psi|^2 \psi \\ \psi(t,x)|_{t=0} = \psi^0(x), \quad \Pi_c \psi^0 = 0 \end{cases}, \quad x \in \mathbb{R}, \ \hbar \ll 1, \tag{13.12}$$

where the actual semiclassical parameter is the splitting ω and the effective nonlinearity parameter is given by η: where

$$\eta = c\frac{\epsilon}{\omega} = O(1) \quad \text{and} \quad \omega = O\left(e^{-\rho/\hbar}\right), \quad \text{as} \ \hbar \to 0,$$

for some $\rho > 0$. We refer to [12] for details.

The proof is organized in three different steps. In the first step we give the global existence of the solution of the time-dependent Schrödinger equation and we prove some useful conservation laws and a priori estimates of the solution. In the second step we give the explicit solution of the two-level approximation obtained by means of the restriction of the complete equation to a bi-dimensional space. Finally, in the third step, we prove the stability of the two-level approximation. Let us emphasize that the first two steps make use of some standard techniques of nonlinear operator theory and of nonlinear dynamical system theory. In contrast, the third step makes use of some particular Gronwall-type estimates on the evolution operator applied to the nonlinear term.

13.5.1 Global existence of the solution and conservation laws

Here, we prove that the Cauchy problem (13.12) admits a solution for all time provided that the strength ϵ of the nonlinear perturbation is small enough. Moreover, we prove an a priori estimate of the solution ψ.

The following results hold.

Theorem 1. *There exist $\hbar^\star > 0$ and $\epsilon_0 > 0$ such that for any $\hbar \in (0, \hbar^\star]$ and $\epsilon \in [-\epsilon_0, \epsilon_0]$, then the Cauchy problem (13.12) admits a unique solution $\psi(t,x) \in H^1$ for any $t \in \mathbb{R}$. Moreover, we have conservation of the norm*

$$\|\psi(t,\cdot)\| = \|\psi^0(\cdot)\| = 1, \tag{13.13}$$

conservation of the energy

$$E(\psi) = \hbar^2 \left\|\frac{\partial \psi}{\partial x}\right\|^2 + \langle V\psi, \psi\rangle + \tfrac{1}{2}\epsilon\|\,\psi^2\,\|^2 = E(\psi^0), \tag{13.14}$$

and the following estimates

$$\|\psi\|_p \le C\hbar^{-\frac{p-2}{4p}}, \ \ \forall p \in [2,+\infty], \quad and \quad \left\|\frac{\partial \psi}{\partial x}\right\| \le C\hbar^{-\frac{1}{2}} \tag{13.15}$$

for some positive constant C independent of t and $\hbar$.

Proof. The first part of the theorem is an immediate consequence of the assumption on ψ^0 where

$$\psi^0 = c_1\varphi_1 + c_2\varphi_2 \in H^1, \quad c_{1,2} = \langle \psi^0, \varphi_{1,2}\rangle,$$

since the two eigenvectors $\varphi_{1,2} \in H^1$. Therefore, existence of the global solution $\psi \in C(\mathrm{R}, H^1)$ and the conservation laws (13.13) and (13.14) follow from known results (see, e.g., [14] and the references therein) for any $\epsilon > 0$ (repulsive nonlinear perturbation) and for any $\epsilon \in (-\epsilon_0, 0)$ for some $\epsilon_0 > 0$ (attractive nonlinear perturbation). The proof of the estimate (13.15) is a consequence of the conservation of the energy functional and from the Gagliardo–Nirenberg inequality. Indeed, the conservation of the energy functional gives that

$$\left\|\frac{\partial \chi}{\partial x}\right\|^2 + \tfrac{1}{2}[\mathrm{sign}(\epsilon)]\|\chi^2\|^2 \le \Lambda\rho^2$$

where

$$\chi = \rho\psi, \quad \rho = \frac{|\epsilon|^{1/2}}{k} \ll 1, \quad k = \frac{\hbar}{\sqrt{2m}}, \quad \Lambda = \frac{E(\psi^0) - V_{min}}{k^2} \sim \hbar^{-1},$$

from which it follows that

$$\left\|\frac{\partial \chi}{\partial x}\right\|^2 \le \rho^2|\Lambda| + \tfrac{1}{2}\|\chi^2\|^2 = \rho^2|\Lambda| + \tfrac{1}{2}\|\chi\|_4^4. \tag{13.16}$$

Now, we make use of the Gagliardo–Nirenberg inequality (see, for instance, [16])

$$\|\chi\|_{2\sigma+2}^{2\sigma+2} \le C\left\|\frac{\partial \chi}{\partial x}\right\|^\sigma \|\chi\|^{2+\sigma}, \ \ \forall \sigma \ge 0, \tag{13.17}$$

which gives, for $\sigma = 1$, the estimate

$$\|\chi\|_4^4 \le C\left\|\frac{\partial \chi}{\partial x}\right\|\|\chi\|^3 \le C\left\|\frac{\partial \chi}{\partial x}\right\|\rho^3$$

since $\|\chi\| = \rho\|\psi\| = \rho$ and $\|\psi\| = 1$. By inserting this inequality in (13.16) it follows that $\left\|\frac{\partial\chi}{\partial x}\right\|$ satisfies

$$\left\|\frac{\partial\chi}{\partial x}\right\|^2 \le \rho^2|\Lambda| + C\rho^3\left\|\frac{\partial\chi}{\partial x}\right\| \tag{13.18}$$

for any $t \in \mathrm{R}$. From (13.18) immediately follows that

$$\left\|\frac{\partial\chi}{\partial x}\right\| \le \sqrt{|\Lambda|}\,\rho\,(1 + o(1))\,, \quad \text{as} \quad \rho \to 0.$$

Hence, $\left\|\frac{\partial\psi}{\partial x}\right\| \le C\sqrt{|\Lambda|}$ and, from (13.17), we have that

$$\|\psi\|_p \le C\left\|\frac{\partial\psi}{\partial x}\right\|^{\sigma/p} \le C|\Lambda|^{(p-2)/4p}$$

where we choose now $\sigma = \frac{p-2}{2}$, i.e., $p = 2\sigma + 2$. $\qquad\square$

13.5.2 Two-level approximation

Let us make the substitution $\psi \to e^{-i\Omega t/\hbar}\psi$ and let us introduce the *slow time* defined in (13.4), hence equation (13.12) takes the form (let us still denote, with abuse of notation, the solution by ψ)

$$i\hbar\psi' = (H - \Omega)\psi + \epsilon|\psi|^2\psi, \quad \psi(x,0) = \psi^0(x), \tag{13.19}$$

where $'$ denotes the derivative with respect to the slow time τ. Let us write the solution of this equation in the form

$$\psi(\tau,x) = a_R(\tau)\varphi_R(x) + a_L(\tau)\varphi_L(x) + \psi_c(\tau,x), \tag{13.20}$$

where $a_R(\tau)$ and $a_L(\tau)$ are unknown complex-valued functions depending on the slow time τ and $\psi_c = \Pi_c\psi$, Π_c is the projection onto the space orthogonal to the two-dimensional space spanned by the two "single-well" states φ_R and φ_L, i.e.,

$$\langle\psi_c, \varphi_R\rangle = \langle\psi_c, \varphi_L\rangle = 0.$$

By substituting ψ by (13.20) in equation (13.19) we obtain that a_R, a_L and ψ_c must satisfy the system of differential equations

$$\begin{cases} ia'_R = f_R, \\ ia'_L = f_L, \\ i\omega\psi'_c = (H - \Omega)\psi_c + \epsilon\Pi_c|\psi|^2\psi, \end{cases} \tag{13.21}$$

where

$$f_R = -a_L + \frac{\epsilon}{\omega}\langle \varphi_R, |\psi|^2\psi\rangle = -a_L + \eta|a_R|^2 a_R + \frac{\epsilon}{\omega}\dot{r}_R, \qquad (13.22)$$

$$f_L = -a_R + \frac{\epsilon}{\omega}\langle \varphi_L, |\psi|^2\psi\rangle = -a_R + \eta|a_L|^2 a_L + \frac{\epsilon}{\omega}r_L \qquad (13.23)$$

and

$$c = \|\varphi_R^2\|^2 = \|\varphi_L^2\|^2 = O(\hbar^{-1}), \qquad (13.24)$$

and where r_R and r_L are given by

$$\begin{aligned}
r_R &= \langle\varphi_R, |\psi|^2\psi\rangle - |a_R^2|a_R\langle\varphi_R, |\varphi_R|^2\varphi_R\rangle \\
&= \langle\varphi_R, |\psi|^2\phi_L\rangle + a_R\langle|\varphi_R|^2, |\phi_L|^2 + a_R\varphi_R\bar{\phi}_L + \bar{a}_R\bar{\varphi}_R\phi_L\rangle, \\
r_L &= \langle\varphi_L, |\psi|^2\psi\rangle - |a_L^2|a_L\langle\varphi_L, |\varphi_L|^2\varphi_L\rangle \\
&= \langle\varphi_L, |\psi|^2\phi_R\rangle + a_L\langle|\varphi_L|^2, |\phi_R|^2 + a_L\varphi_L\bar{\phi}_R + \bar{a}_L\bar{\varphi}_L\phi_R\rangle,
\end{aligned}$$

where

$$\phi_L = a_L\varphi_L + \psi_c \quad \text{and} \quad \phi_R = a_R\varphi_R + \psi_c\,.$$

We denote by *two-level approximation* the solutions b_R and b_L of the system of ordinary differential equations

$$\begin{cases} ib_R' = -b_L + \eta|b_R|^2 b_R \\ ib_L' = -b_R + \eta|b_L|^2 b_L \end{cases}, \quad b_{R,L}(0) = a_{R,L}(0), \ \eta = \frac{\epsilon c}{\eta}, \qquad (13.25)$$

obtained by neglecting the remainder terms r_R and r_L in (13.22) and (13.23). It is easy to see that the solution of this system satisfies the conservation law

$$\begin{aligned}
|b_R(\tau)|^2 + |b_L(\tau)|^2 &= |b_R(0)|^2 + |b_L(0)|^2 \\
&= |a_R(0)|^2 + |a_L(0)|^2 = 1,
\end{aligned} \qquad (13.26)$$

and, moreover, it is also possible to explicitly compute the solution of (13.25) by means of elliptic functions. In particular, let us introduce the imbalance function and the relative phase, respectively defined as

$$\begin{aligned}
z(\tau) &= |b_R(\tau)|^2 - |b_L(\tau)|^2 \\
\theta(\tau) &= \arg[b_R(\tau)] - \arg[b_L(\tau)]\,.
\end{aligned} \qquad (13.27)$$

Then it follows that these two functions have to satisfy the system of ordinary differential equations

$$\begin{cases} z' = 2\sqrt{1 - z^2}\sin\theta \\ \theta' = -\dfrac{2z}{\sqrt{1 - z^2}}\cos\theta - \eta z, \end{cases}$$

which admits the conservation law

$$I = \sqrt{1 - z^2}\cos\theta - \tfrac{1}{4}\eta z^2 \,. \tag{13.28}$$

Hence, $z(\tau)$ satisfies the differential equation of the first order

$$(z')^2 = 4(1 - z^2) - 4\left(I + \tfrac{1}{4}\eta z^2\right)^2 ,$$

solution of which is given by means of the elliptic Jacobian functions (see §7.10 [3]):

$$z(\tau) = \begin{cases} A\,\mathrm{cn}\left[A\eta(\tau - \tau_0)/2k, k\right], & \text{if } k < 1, \\ A\,\mathrm{dn}\left[A\eta(\tau - \tau_0)/2, 1/k\right], & \text{if } k > 1, \end{cases}$$

where τ_0 depends on the initial condition,

$$A = \frac{2\sqrt{2}}{\eta}\left[\sqrt{\tfrac{1}{4}\eta^2 + 1 + I\eta} - \left(1 + \tfrac{1}{2}I\eta\right)\right]^{1/2} ,$$

and

$$k^2 = \tfrac{1}{2}\left[1 - \frac{1 + \tfrac{1}{2}I\eta}{\sqrt{\tfrac{1}{4}\eta^2 + 1 + I\eta}}\right] . \tag{13.29}$$

We emphasize that $z(\tau)$ periodically assumes positive and negative values if, and only if, $k < 1$.

13.5.3 Stability of the two-level approximation

Our main result consists in proving the stability of the two-level approximation when we restore the remainder terms r_R and r_L in equation (13.25).

We prove that:

Theorem 2. *Let $\psi_c = \Pi_c\psi$, $a_R(\tau) = \langle\psi,\varphi_R\rangle$ and $a_L(\tau) = \langle\psi,\varphi_L\rangle$, where ψ is the solution of equation (13.19), let $b_R(\tau)$ and $b_L(\tau)$ be the solution of the system of ordinary differential equations (13.25). Let ϵ satisfy the condition (13.9). Then, for any $\tau' > 0$ there exists a positive constant C independent of ϵ, $\hbar$ and τ such that:*

$$|b_{R,L}(\tau) - a_{R,L}(\tau)| \le Ce^{-C\hbar^{-1}} \quad \text{and} \quad \|\psi_c(\cdot,\tau)\| \le Ce^{-C\hbar^{-1}} \tag{13.30}$$

for any $\hbar \in (0, \hbar^\star]$ and for any $\tau \in [0, \tau']$.

Proof. For the sake of simplicity, hereafter, let us drop the parameters where this does not cause misunderstanding. The equations (13.21) and (13.25) can be written as

$$A' = F(A) + R \quad \text{and} \quad B' = F(B), \quad A(0) = B(0) = a(0), \tag{13.31}$$

where $A, B \in S^2$, $S^2 = \{(z_1, z_2) \in \mathbf{C}^2 : |z_1|^2 + |z_2|^2 \leq 1\}$, are defined as

$$A = \begin{pmatrix} a_R \\ a_L \end{pmatrix}, \quad B = \begin{pmatrix} b_R \\ b_L \end{pmatrix}, \quad R = -i\frac{\epsilon}{\omega}\begin{pmatrix} r_R \\ r_L \end{pmatrix},$$

$$F(A) = -i\begin{pmatrix} -a_L + \eta|a_R|^2 a_R \\ -a_R + \eta|a_L|^2 a_L \end{pmatrix}.$$

By definition it immediately follows that the function $F : S^2 \to \mathbf{C}^2$ satisfies the Lipschitz condition:

$$|F(A) - F(B)| \leq L|A - B|, \quad L = 1 + 3\eta. \tag{13.32}$$

We give now the following estimate of the norm of the remainder term ψ_c. Let

$$\beta = \max[c\epsilon, \omega] = \omega\max[1, \eta] \sim e^{-\rho/\hbar}$$

where η is the nonlinearity parameter defined in (13.9). Let $\psi_c = \Pi_c\psi$ where ψ is the solution of equation (13.19); it satisfies the following uniform estimate

$$\|\psi_c\| \leq C\beta\hbar^{-3/2}\left[\exp[C\tau\hbar^{-1/2}] + 1\right], \quad \forall\tau \in \mathbf{R}, \tag{13.33}$$

for some positive constant C independent of $\hbar$, ϵ and t.

We emphasize that here we actually make use of the essential assumption on the dimension $d = 1$; for the proof of this result we refer to [12].

From this inequality, by the definition of the remainder terms $r_{R,L}$ and from the fact that any vector φ_R and φ_L is practically localized on just one well, it immediately follows that for any fixed $\tau' > 0$ there exists $C > 0$ such that

$$\max\left[|r_R|, |r_L|\right] \leq C\beta\hbar^{-2}e^{C\hbar^{-1/2}}, \quad \forall\tau \in [0, \tau'].$$

In fact, let us only consider the term $|r_R|$, the other term $|r_L|$ could be treated in the same way. By definition and since $\max[|a_R|, |a_L|] \leq 1$, it follows that

$$|r_R| \leq + \left|\langle\varphi_R\bar{\varphi}_L, |\psi|^2\rangle\right| \tag{13.34}$$

$$+ \left|\langle\varphi_R|\psi|^2, \psi_c\rangle\right| \tag{13.35}$$

$$+ \left|\langle|\varphi_R|^2, |\phi_L|^2 + a_R\varphi_R\bar{\phi}_L + \bar{a}_R\bar{\varphi}_R\phi_L\rangle\right| \tag{13.36}$$

and we estimate separately each term. Indeed, one has that (for the details see [12] again)

$$\left|\langle\varphi_R\varphi_L, |\psi|^2\rangle\right| \leq \|\varphi_R\bar{\varphi}_L\|_\infty \cdot \|\psi^2\|_1 \leq C\omega,$$

$$\left|\langle\varphi_R|\psi|^2, \psi_c\rangle\right| \leq \|\varphi_R\|_\infty \cdot \|\psi^2\| \cdot \|\psi_c\| \leq C\beta\hbar^{-2}e^{C\hbar^{-1/2}}$$

and

$$\left| \langle |\varphi_R|^2, |\phi_L|^2 + a_R\varphi_R\bar{\phi}_L + \bar{a}_R\bar{\varphi}_R\phi_L \rangle \right|$$
$$\leq C \left[\|\varphi_R\varphi_L\|_\infty + \|\varphi_R^2\|_\infty \|\psi_c\|^2 + \|\varphi_R\varphi_L\|_\infty \|\psi_c\| \right] \leq C\omega .$$

Collecting all these results we finally obtain the estimate of the remainder term r_R.

The proof of the theorem is almost done. Indeed, equations (13.31) can be rewritten in the integral form:

$$A(\tau) = A(0) + \int_0^\tau F[A(s)]ds + \int_0^\tau R ds$$

and

$$B(\tau) = B(0) + \int_0^\tau F[B(s)]ds,$$

from which it follows that for any $\tau \in [0, \tau']$,

$$|A(\tau) - B(\tau)| \leq \int_0^\tau |F[A(s)] - F[B(s)]| \, ds + \int_0^\tau |R| ds$$
$$\leq a \int_0^\tau |A(s) - B(s)| \, ds + b\tau, \quad a = L, \quad b = C \frac{\epsilon\beta\hbar^{-2}e^{C\hbar^{-1/2}}}{\omega}$$

from the previous estimates. From this inequality and by means of Gronwall's lemma [7] we finally obtain that

$$|A(\tau) - B(\tau)| \leq \frac{C}{L} \frac{\epsilon\beta\hbar^{-2}e^{C\hbar^{-1/2}}}{\omega}$$

proving (13.30), since

$$\frac{\omega + \epsilon}{C'\omega} \leq L = 1 + 3\eta \leq C' \frac{\omega + \epsilon}{\omega},$$

for some $C' > 0$, which implies that $\frac{\beta}{L\omega} \leq C$ for some $C > 0$. $\qquad \square$

13.6 Destruction of the beating motion for large nonlinearity

Let us consider the motion of the *center of mass* defined here as

$$\langle g \rangle^\tau = \langle \psi, g\psi \rangle = \int_{\mathbf{R}} g(x)|\psi(\tau, x)|^2 dx$$

where $g \in C(R) \cap L^2(\mathbf{R})$ is a given bounded function such that $g(-x) = -g(x)$. We have that

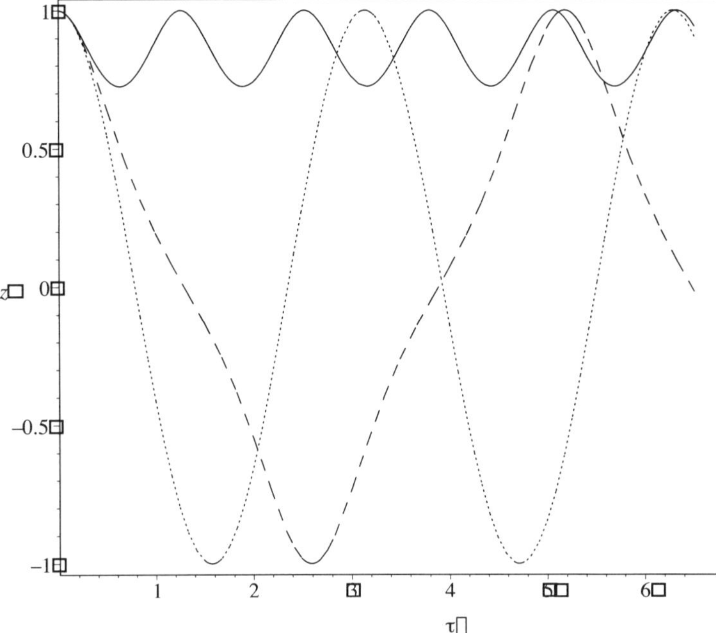

Fig. 13.1. Destruction of the beating motion of the *center of mass* for nonlinearity larger than a critical value. Here, we plot the *imbalance function* $z(\tau)$ for different values of the nonlinearity parameter η. For $\eta = 0$ (point line) and $\eta = 3.8$ (broken line) we still have a beating motion; in contrast for η larger than the critical value 4, e.g., $\eta = 6.5$ (full line), the beating motion is forbidden.

$$\langle g \rangle^\tau = g_0[|a_R(\tau)|^2 - |a_L(\tau)|^2] + r, \quad g_0 = \langle \varphi_R, g\varphi_R \rangle,$$

where the remainder term r satisfies the uniform estimate

$$|r| = 2\,|\Re\left[a_R \bar{a}_L \langle X\varphi_R, \varphi_L \rangle + \langle X\psi, \psi_c \rangle \right]|$$
$$\leq 2\left[\|\varphi_R \varphi_L\|_\infty + \|X\|_\infty \|\psi\| \|\psi_c\|\right]$$
$$\leq Ce^{-C\hbar^{-1}}, \quad \forall \tau \in [0, \tau'].$$

If we denote by $z(\tau)$ the imbalance function defined in (13.29), then in the semiclassical limit it follows that

$$|a_R(\tau)|^2 - |a_L(\tau)|^2 \sim z(\tau), \quad \forall \tau \in [0, \tau'],$$

hence

$$\langle g \rangle^\tau \sim g_0 z(\tau), \quad \forall \tau \in [0, \tau'].$$

Then we have that:

Theorem 3. *Let k^2 be defined as in (13.29), which depends on the initial wavefunction ψ^0. For any $\tau' > 0$ fixed and in the semiclassical limit, the function $\langle X \rangle^\tau$ is, up to a small remainder term, a periodic function for any $\tau \in [0, \tau']$. In particular, if:*

i) $k^2 < 1$, then $\langle g \rangle^\tau$ periodically assumes positive and negative values (i.e., the beating motion still persists);

ii) $k^2 > 1$, then $\langle g \rangle^\tau$ has a definite sign (i.e., the beating motion is forbidden).

Remark. Let us close by emphsizing that when the wavefunction is initially prepared on just one well, e.g., $\psi^0 = \varphi_R$, then

$$I = -\tfrac{1}{4}\eta \quad \text{and} \quad k^2 = \tfrac{1}{16}\eta^2.$$

Therefore, from the theorem above it follows that for $|\eta|$ larger than the critical value 4, the beating motion is forbidden (see, for instance, Fig. 13.1, for different values of η: $\eta = 0$, 3.8 and 6.5) proving thereby the proposition.

References

1. F. Dalfovo, S. Giorgini, L.P. Pitavskii, S. Stringari, Theory of Bose–Einstein condensation in trapped gases. *Rev. Mod. Phys.* **71** (1999), 463–512.
2. E.B. Davies, Symmetry breaking for a non-linear Schrödinger operator *Comm. Math. Phys.* **64** (1979), 191–210.
3. H.T. Davis, Introduction to nonlinear differential and integral equations, Dover Publ., New York, 1962.
4. J. Ginibre, G. Velo, On a class of nonlinear Schrödinger equations. I. The Cauchy problem, general case, *J. Func. Anal.* **32** (1979), 1–32.
5. V. Grecchi, A. Martinez, Non-linear Stark effect and molecular localization, *Comm. Math. Phys.* **166** (1995), 533–48.
6. V. Grecchi, A. Martinez, A. Sacchetti, Destruction of the beating effect for a non-linear Schrödinger equation. *Comm. Math. Phys.* **227** (2002), 191–209.
7. E. Hille, Lectures on Ordinary Differential Equations, Addison-Wesley, London, 1969.
8. G. Jona-Lasinio, C. Presilla, C. Toninelli, Interaction induced localization in a gas of pyramidal molecules, *Phys. Rev. Lett.* **88** (2002), 123001.
9. C. Presilla, Interaction induced localization in a gas of pyramidal molecules. Proceedings of the conference *Multiscale Methods in Quantum Mechanics: Theory and Experiment,* Accademia Nazionale dei Lincei, Roma 16–20 December 2002. See G. Jona-Lasinio, C. Presilla, C. Toninelli, Chapter 10, this volume, pp. 119–27.
10. S. Raghavan, A. Smerzi, S. Fantoni, S.R. Shenoy, Coherent oscillations between two weakly coupled Bose–Einstein condensates: Josephson effects, π oscillations, and macroscopic quantum self-trapping, *Phys. Rev. A* **59**, 620–33 (1999).
11. A. Sacchetti, Nonlinear time-dependent Schrödinger equations: the Gross—Pitaevskii equation with double-well potential, preprint mp-arc 02-208 (2002). To be published in *J. Evolution Equations.*

12. A. Sacchetti, Nonlinear time-dependent one-dimensional Schrödinger equation with double well potential. *SIAM J. Math. Anal.* **35** (2004), 1160–76.
13. C. Sulem, P.L. Sulem, *The Nonlinear Schrödinger Equation: Self-focusing and Wave Collapse*, Springer-Verlag, New York, 1999.
14. A. Soffer, M.I. Weinstein, Multichannel nonlinear scattering for nonintegrable equations, *Comm. Math. Phys.* **133** (1990), 119–46.
15. A. Vardi, On the role of intermolecular interactions in establishing chiral stability, *J. Chem. Phys.* **112** (2000), 8743.
16. M.I. Weinstein, Nonlinear Schrödinger equations and sharp interpolation estimates, *Comm. Math. Phys.* **87** (1983), 567–76.
17. R. Weder, $L^p - L^{p'}$ estimates for the Schrödinger equation on the line and inverse scattering for the nonlinear Schrödinger equation with a potential, *J. Funct. Anal.* **170** (2000), 37–68.

14

Classical and Quantum: Some Mutual Clarifications

V. Scarani

Group of Applied Physics
University of Geneva
20, rue de l'Ecole-de-Médecine
CH-1211 Geneva 4, Switzerland

Summary. This paper presents two unconventional links between quantum and classical physics. The *first link* appears in the study of quantum cryptography. In the presence of a spy, the quantum correlations shared by Alice and Bob are imperfect. One can either process the quantum information, recover perfect correlations and finally measure the quantum systems; or, one can perform the measurements first and then process the classical information. These two procedures tolerate exactly the same error rate for a wide class of attacks by the spy. The *second link* is drawn between the quantum notions of "no-cloning theorem" and "weak-measurements with post-selection", and simple experiments using classical polarized light and ordinary telecom devices.

14.1 Introduction

The boundary between classical and quantum physics is a fascinating region, that in my opinion, in spite of several important explorations, has not delivered its deepest treasures. I will try to motivate this optimistic view on the future of research in physics by presenting some remarkable links between "quantum" and "classical" physics.

We have often read in old textbooks or popular books that quantum physics is the physics of the "infinitely small", while the "everyday world" is governed by classical physics. This might be considered, and probably is, a very naive view. Bohr maintained that the distinction between the classical measurement device and the quantum measured system is arbitrary but is necessary for our understanding. The current view of the physicists working in the field, is that *everything is quantum*, the classicality emerging through interactions (the "everyday world" appears then to be classical because of the huge amount of interacting particles involved). This last view, the emergence of classical behavior simply because of interaction, is nowadays unchallenged by observation: no phenomenon can be produced as an evidence of its falseness. Thus, it is a satisfactory description for any practical purpose, although one may question its validity as a *Weltanschauung*.

The links between "classical" and "quantum" that am I going to present here are of a different nature: they do not seem to arise simply from many interacting quantum objects that together exhibit classical behavior. The first link (Section 14.2) deals with quantum cryptography[1]. The second link (Section 14.3) shows how typical "quantum" notions (namely, the no-cloning theorem and the idea of weak measurements with post-selection) manifest themselves in phenomena that can be described using an entirely classical theory of light, and that can be revealed using the devices of ordinary telecommunication networks.

14.2 Classical bounds in quantum cryptography

Quantum cryptography is nowadays the most developed application of quantum information theory [9]. A more exact name for quantum cryptography would be *quantum key distribution* (QKD): the goal of the quantum processing is to establish a *secret key* between two distant partners, Alice and Bob, avoiding the attacks of a possible eavesdropper Eve. Once a common secret key is established, Alice and Bob will encode the message using classical secret-key protocols, known to be unbreakable even if the message is sent on a public authenticated channel.

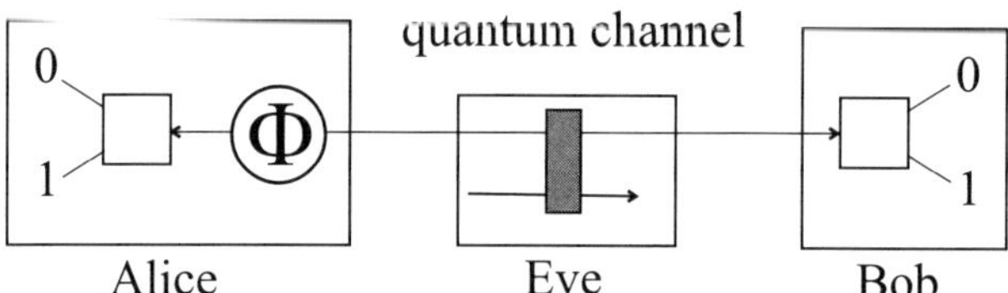

Fig. 14.1. The scheme of the QKD implementation with entangled states. Alice prepares a maximally entangled state $|\Phi\rangle$. She measures one particle (here, a two-level system) and forwards the other one to Bob. The spy Eve accesses the quantum channel and tries to obtain some information by interacting with the flying particle.

We describe (Fig. 14.1) an implementation of QKD that uses a source of entangled states, and for clarity we speak of two-dimensional quantum systems (qubits). Alice has a source that produces a pair of qubits in the maximally entangled state

---

[1] This was the topic of my talk during the Workshop *Multiscale Methods in Quantum Mechanics*, Rome, 16–20 December, 2002.

$$|\Phi\rangle_{AB} = \frac{1}{\sqrt{2}}\left(|+z\rangle \otimes |+z\rangle + |-z\rangle \otimes |-z\rangle\right)$$

$$= \frac{1}{\sqrt{2}}\left(|+x\rangle \otimes |+x\rangle - |-x\rangle \otimes |-x\rangle\right). \qquad (14.1)$$

She keeps one qubit and forwards the other one to Bob. In the absence of Eve:
(i) if Alice and Bob measure the same observable, either σ_z or σ_x, they obtain
the same result, the same *random bit*; (ii) if one of the partners measures
σ_z and the other σ_x, they obtain completely uncorrelated random bits. This
protocol is repeated a large number of times. At the end, the items in which
Alice and Bob have performed different measurements are discarded later by
public communication on the classical channel, leaving Alice and Bob with a
list of perfectly correlated random bits: the secret key. This is what happens
in the absence of the spy.

Eve can in principle do whatever she wants on the quantum channel. The
security of QKD comes from the fact that, since any measurement or interaction perturbs the state, Eve's intervention cannot pass unnoticed: Alice and
Bob know that someone is spying. Two situations are then possible. (I) Eve
has got a "small" amount of information; in this case, Alice and Bob can
process their data in order to obtain a shorter but completely secret key. Such
classical protocols are the object of important studies in classical information
theory. (II) Eve has got "too much" information; then Alice and Bob discard
the whole key. This may seem a failure, but it is not: it simply means that
the spy has no other alternative than cutting the channel and forbid any communication; and this *achieves* the goal of cryptography, because no encrypted
message is ever sent that the spy could decode.

It is then important to quantify the words "small" and "too much" in the
previous discussion: what is the amount of Eve's information that Alice and
Bob can tolerate, that is, at what critical value are they obliged to discard
the whole key? Here is where remarkable links appear between classical and
quantum information.

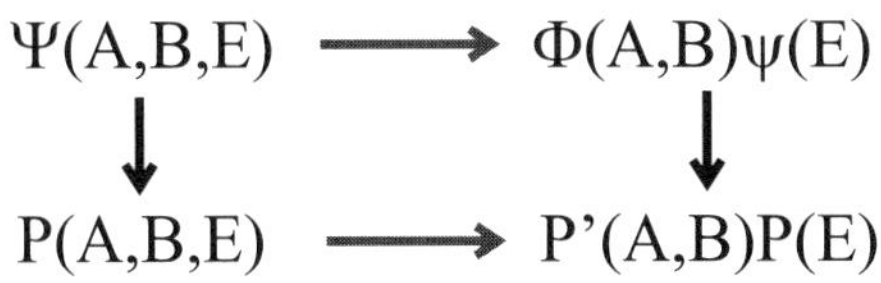

$$\Psi(A,B,E) \longrightarrow \Phi(A,B)\psi(E)$$
$$\downarrow \qquad\qquad\qquad \downarrow$$
$$P(A,B,E) \longrightarrow P'(A,B)P(E)$$

Fig. 14.2. Possible ways for the extraction of a secret key. One starts from a global
quantum state $\Psi(A, B, E)$ of Alice, Bob and Eve, and wants to end up with a classical secret key $P'(A, B)P(E)$ with $P'(A = B) = 1$. Horizontal arrows: distillation,
quantum or classical; vertical arrows: measurement of the quantum system, leading
to a classical probability distribution.

We refer to Fig. 14.2. Because of Eve's intervention, before any measurement the quantum system is in a three-party entangled state of Alice-Bob-Eve, Ψ_{ABE}. Alice and Bob on their own share the mixed state ρ_{AB}, obtained from Ψ_{ABE} by partial trace on Eve's system. Two procedures are then possible:

(a) The one that we described above: all the partners make a measurement, ending in a classical probability distribution $P(A, B, E)$. Then, Alice and Bob apply classical protocols (*advantage distillation*) in order to extract a shorter secret key, that is, a shorter list of bits distributed according to a new distribution $P'(A, B)P'(E)$ in which Eve is uncorrelated and $P'(A = B) = 1$.

(b) If the state ρ_{AB} is entangled, Alice and Bob can delay any measurement and process many copies of ρ_{AB}, to obtain a smaller number of copies of $|\Phi\rangle_{AB}$ — and in this case, automatically Eve is uncorrelated. This procedure is known as *entanglement distillation*, and is one of the fundamental processes of quantum information. Once Alice and Bob have $|\Phi\rangle_{AB}$, the measurement provides them immediately with the secret key.

Having understood this, we can state the main results that have been obtained:

- Classical advantage distillation of $P(A, B, E)$ is possible for bits if and only if quantum entanglement distillation is possible for the state ρ_{AB} (which is equivalent to asking that ρ_{AB} be entangled in the case of qubits). This was demonstrated by Gisin and Wolf when Eve uses the so-called optimal individual attack [10], and has been recently extended to all individual attacks [1].
- The same holds for dits (d-valued random variables) and qudits (d-dimensional quantum systems), under Eve's individual attack that is supposed to be the optimal one [2, 6]. The demonstration is more involved because not all entangled states of two qudits are distillable.
- If ρ_{AB} is entangled enough to violate a Bell inequality, then a secret key can be extracted from $P(A, B, E)$ in an *efficient* way, that is, using only one-way communication. This was first proved in Ref. [8]; for the state-of-the-question, see [2]. A similar result holds for a multi-partite scheme of key distribution known as "quantum secret sharing" [11].

Mainly because Eve's optimal attack is not generally known, there are still several open questions. The most important ones are reviewed in the last section of Ref. [2].

This concludes my first "unconventional" link between the classical and the quantum worlds: at the level of information processing, specifically of the extraction of a secret key from an initially noisy distribution/state, the critical parameters are exactly the same, irrespective whether the purification of the correlations is performed at the quantum or at the classical level. Moreover, a typically quantum feature such as the violation of Bell's inequalities is related to the efficiency of the classical key-extraction procedure.

14.3 Quantum physicists meet telecom engineers

This section is devoted to another kind of unconventional link between the classical and the quantum world. I prefer to let the examples speak first and draw my conclusions later.

14.3.1 No-cloning theorem

The first example concerns the no-cloning theorem, a well-known primitive concept of quantum information [14]. In its basic form, it states that no evolution (or more generally, no trace-preserving completely positive map) can bring $|\psi\rangle \otimes |0\rangle$ onto $|\psi\rangle \otimes |\psi\rangle$ for an unknown state $|\psi\rangle$.

This no-go theorem has motivated the search for an *optimal quantum cloner*: given that perfect cloning is impossible, what is the best one can do? Optimal cloners have indeed been found and widely studied; all the meaningful references can be found in any basic text on quantum information, e.g., [4]. In the course of these investigations, a sharp link has been found between optimal cloning and the well-known phenomenon of amplification of light [12]: stimulated emission of light in a given mode (perfect amplification, or cloning) cannot be done without spontaneous emission (random amplification). Suppose that N photons enter an amplifier, and at the output one selects the cases in which exactly $M > N$ photons are found: it turns out that this process *realizes the optimal quantum cloning* from N to M copies. The *fidelity* of the amplification is the ratio between the mean number of photons found in the initial mode (i.e., the mean number of correct copies) and the total number of copies, M here. The optimal fidelity is found to be

$$\mathcal{F}^{opt}_{N \to M} = \frac{MN + M + N}{M(N + 2)} \,.\tag{14.2}$$

We realized an experimental demonstration of optimal cloning using the principle just described to clone the polarization of light [7]. Polarized light of intensity μ_{in} is sent into a conventional fiber amplifier (exactly as those that are used in telecommunications); at the output, we have an intensity μ_{out}; we separate the input polarization mode from its orthogonal, and measure the fidelity. The theoretical prediction for this experiment is

$$\bar{\mathcal{F}}_{\mu_{in} \to \mu_{out}} = \frac{Q\mu_{out}\,\mu_{in} + \mu_{out} + \mu_{in}}{Q\mu_{out}\mu_{in} + 2\mu_{out}}\tag{14.3}$$

where $Q \in [0, 1]$ is a parameter related to the quality of the amplification process. The experimental results are in excellent agreement (Fig. 14.3).

It is striking to notice that for $Q = 1$, formula (14.3) is exactly the same as (14.2). Its meaning is however rather different. In our experiment, *everything is classical*: the laser light is in a coherent state, therefore it can be described by a classical field; the amplifier is "classical" in the sense that it transforms coherent states into coherent states. The quantities μ_{in} and μ_{out} that appear in eq.(14.3) are not photon numbers as the N and M of eq. (14.2), but mean values, that have been measured using an intensity detector. As I said above, all the devices (source, fibers, amplifier, detectors) are typical of telecom engineering.

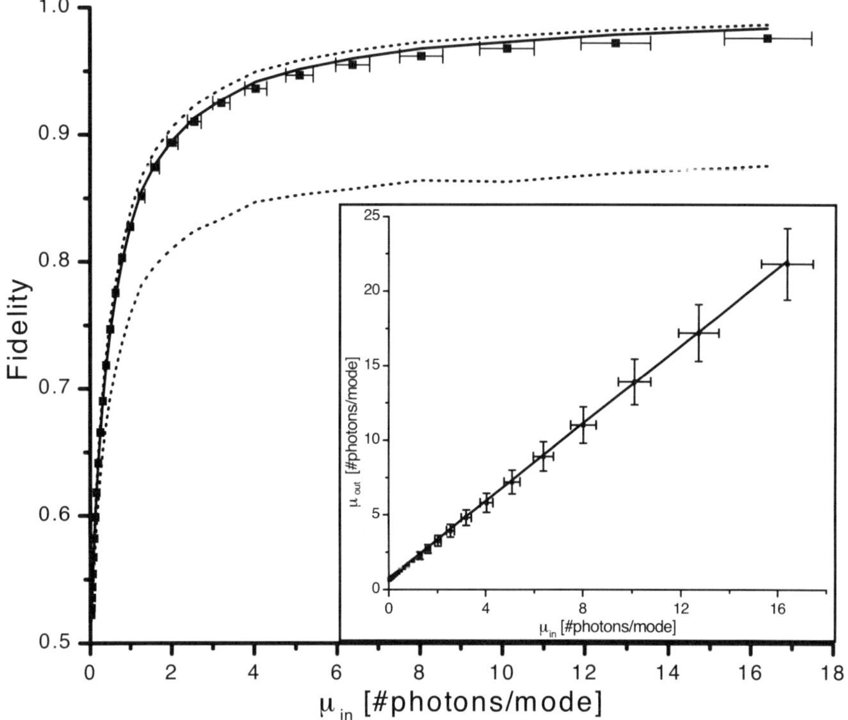

Fig. 14.3. Inset: μ_{out} as a function of μ_{in}; the linear fit shows that we are far from the saturation of the amplifier. Main figure: fidelity as a function of μ_{in}. Solid line: $Q = 0.8$, best fit with eq. (14.3). Dotted lines: upper: $Q = 1$ (optimal cloning); lower: $Q = 0$ (no cloning). From Ref. [7].

14.3.2 Weak measurements with post-selection

The second example is related to the meaning and physics of the *measurement process*, a widely debated topic of the foundations of quantum mechanics. In this context, Aharonov, Vaidman and others introduced the notion of *weak measurement with post-selection* [3], sometimes called the "two-state formalism" of quantum mechanics. The authors' intention in studying this formalism is strongly motivated by interpretational issues; that is why most physicists tend to look at these concepts as artificial ones, introduced on purpose, and that do not add anything to physics itself. To date, apart from some experiments that were designed on purpose, only some complex tunnelling phenomena and a puzzling result of a cavity-QED experiment [13] had received clarification through this formalism.

We have found however [5] that this formalism does apply to something that exists and is extremely widespread: once again, the optical telecommunication network. Telecom engineers are performing weak measurements with post-selection in basically all that they do! A modern optical network is composed of different devices connected through optical fibers. With respect to polarization, two main physical effects are present. The first one is *polarization-mode dispersion* (PMD): due to birefringency, different polarization modes propagate with different velocities; in particular, the fastest and the slowest polarization modes are orthogonal. PMD is the most important polarization effect in the fibers. The second effect is *polarization-dependent loss* (PDL), that is, different polarization modes are differently attenuated. PDL is negligible in fibers, but is important in devices like amplifiers, wavelength-division multiplexing couplers, isolators, circulators etc.

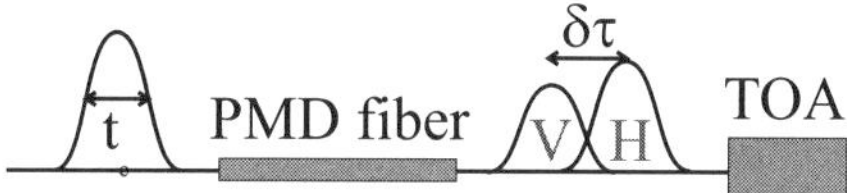

Fig. 14.4. When a polarized pulse passes through a PMD fiber, the polarization mode H (parallel to the birefringency axis in the Poincaré sphere) and its orthogonal V are separated in time. A measurement of the time-of-arrival (TOA) is a measurement, strong or weak, of the polarization.

The first piece of the connection we want to point out is the following: *a PMD element performs a measurement of polarization on light pulses* (Fig. 14.4). In fact, PMD leads to a separation $\delta\tau$ of two orthogonal polarization modes in time. If $\delta\tau$ is larger than the pulse width t_c, the measurement of the time of arrival is equivalent to the measurement of polarization — PMD acts then as a "temporal polarizing beam-splitter". However, in the usual telecom regime $\delta\tau$ is much *smaller* than t_c. In this case, the time of arrival does not achieve a complete discrimination between two orthogonal polarization modes anymore; but still, some information about the polarization of the input pulse is encoded in the modified temporal shape of the output pulse. We are in a regime of *weak measurement* of the polarization. The formulae introduced by Aharonov and co-workers are recovered by measuring the *mean time of arrival*, that is, the "center of mass" of the output pulse.

The second piece of the connection defines the role of PDL: *a PDL element performs a post-selection of some polarization modes*. Far from being an artificial ingredient, post-selection of some modes is the most natural situation in the presence of losses: one does always post-select those photons that have not been lost! This would be trivial physics if the losses were independent of any degree of freedom, just like random scattering; but in the case of PDL, the amount of losses depends on the meaningful degree of freedom, polarization. An infinite PDL would correspond to the post-selection of a precise polarization mode (a pure state, in the quantum language); a finite PDL corresponds to post-selecting different modes with different probabilities (a mixed quantum state).

In summary: by tuning the PMD, we can move from weak to strong measurements of polarization; the PDL performs the post-selection of a pure or of a mixed state of polarization. Any telecom network, devices connected by fibers, is performing "weak measurements with post-selection". Just as in the example of quantum cloning discussed above, all this can be (and *is actually*) described by the classical theory of light.

14.3.3 The fundamental role of entanglement

We have shown that two results thought to be "typically quantum", namely the no-cloning theorem and the theory of weak measurements, can be demonstrated with classical light and standard telecom devices. The key for a deep understanding is the conceptual distinction between *two superposition principles*: the classical one, which is dynamical (fields superpose because Maxwell's equations are linear); and the quantum one, which is kinematical: states are superposed. These two superposition principles, at the level of interpretation, have a completely different meaning. However, it may difficult to tell which is acting in a real situation.

I would like to extend this observation to stress the fundamental role of *entanglement*. In the traditional textbooks of quantum mechanics, entanglement has been considered a kind of a side-issue, and in any case a derived notion: if a composed quantum system is described by a tensor product of Hilbert spaces, and if the superposition principle has to hold in this total space, then non-factorizable states must appear. In other words, traditionally one starts with the quantum physics of the single system, states the superposition principle in this context, and derives the existence of entanglement a posteriori. While this may be an unavoidable approach for a didactic course, I don't think that the view so conveyed is really the whole story. Students meeting the Stern–Gerlach experiment in their "quantum physics" course fail to realize that they have studied its analog with light polarization some months before, in their lectures on "*classical* electrodynamics". But what does it mean? Is a spin $\frac{1}{2}$ classical? Or is polarization a quantum intruder in the classical theory of light?

The solution comes by noticing that the Stern–Gerlach experiment is not the only experimental result involving the spin! The spins of the electrons explain the Mendeleev table via the Pauli exclusion principle, a principle that has no classical analogue; different cross-sections have been observed in scattering experiments, according to whether the full spin was in a symmetric or in the anti-symmetric state; spins couple coherently to one another in nuclear magnetic resonance, or to the polarization of photons in atomic physics, etc. The list may rapidly become very large. But if we give a second glance to this list, we notice that it contains only phenomena in which *two or more* quantum systems are involved. And if we finally notice that "coherent interaction" means "entanglement", we have the solution: *we know that a single spin $\frac{1}{2}$ is a quantum object because we observe the consequences of its entanglement with other spins or other degrees of freedom.* The same can be said for polarization.

The difference between the "classical superposition principle of waves" and the "quantum superposition principle of states" lies in the fact that only the second gives rise to entanglement. If we had only the quantum physics of the single particle (the Stern–Gerlach experiment, Young's double-slit, etc.), the most economic solution would be to adopt once for all the de Broglie–Bohm view of a real particle guided by a hidden wave — and we'd lose all the fascinating view of the world that is inspired by quantum physics.

14.4 Conclusion

The main message I wanted to convey is that "classical" and "quantum" physics — or information — are tightly connected. Specifically, I have discussed how in the analysis of the security of quantum cryptography, we discover numbers that come from the analysis of the security of classical cryptography (Section 14.2); and how experiment with classical light and standard telecom devices can provide demonstrations of the no-cloning theorem and of the theory of weak measurements with post-selection (Section 14.3).

In this text, I reported on results obtained at the University of Geneva under the direction of Prof. Nicolas Gisin, together with Antonio Acín, Nicolas Brunner, Daniel Collins, Sylvain Fasel, Grégoire Ribordy and Hugo Zbinden. I also benefited from several discussions with François Reuse and Antoine Suarez.

References

1. A. Acín, N. Gisin, L. Masanes, quant-ph/0303053, accepted in *Phys. Rev. Lett.*
2. A. Acín, N. Gisin, V. Scarani, quant-ph/0303009, accepted in *Quant. Inf. Comput.*
3. Y. Aharonov, D. Albert, L. Vaidman, *Phys. Rev. Lett.* **60** (1988), 1351. Two recent reviews: Y. Aharonov, L. Vaidman, quant-ph/0105101 (2001), published in *Time in Quantum Mechanics*, J. G. Muga, R. Sala Mayato and I. L. Egusquiza, eds., Lecture Notes in Physics, Springer Verlag, 2002; and A.M. Steinberg, quant-ph/0302003 (2003).
4. D. Bouwmeester, A. Ekert, A. Zeilinger (eds.), *The Physics of Quantum Information*, Springer-Verlag, Berlin, 2000.

5. N. Brunner, A. Acín, D. Collins, N. Gisin, V. Scarani, quant-ph/0306108, accepted in *Phys. Rev. Lett.*; N. Brunner, quant-ph/0309055.

6. D. Bruss, M. Christandl, A. Ekert, B.-G. Englert, D. Kaszlikowski, C. Macchiavello, *Phys. Rev. Lett.* **91** (2003), 097901.

7. S. Fasel, N. Gisin, G. Ribordy, V. Scarani, H. Zbinden, *Phys. Rev. Lett.* **89** (2002), 107901.

8. C. Fuchs, N. Gisin, R.B. Griffiths, C.-S. Niu, *A. Peres, Phys. Rev. A* **56** (1997), 1163.

9. N. Gisin, G. Ribordy, W. Tittel, H. Zbinden, *Rev. Mod. Phys* **74** (2002), 145.

10. N. Gisin and S. Wolf, *Phys. Rev. Lett.* **83** (1999), 4200.

11. V. Scarani, N. Gisin, *Phys. Rev. Lett.* **87** (2001), 117901; idem, *Phys. Rev. A* **65** (2002), 012311.

12. C. Simon, G. Weihs, A. Zeilinger, *Phys. Rev. Lett.* **84** (2000), 2993; J. Kempe, C. Simon, G. Weihs, *Phys. Rev. A* **62** (2000), 032302.

13. H. Wiseman, *Phys. Rev. A* **65** (2002), 032111.

14. W.K. Wootters, W.H. Zurek, *Nature* **299** (1982), 802; P.W. Milonni, M.L. Hardies, *Phys. Lett. A* **92** (1982), 321.

Localization and Delocalization for Nonstationary Models

P. Stollmann

Fakultät für Mathematik
Technische Universität Chemnitz
D-09107 Chemnitz, Germany

15.1 Introduction: Leaving stationarity

In recent years there has been considerable progress concerning mathematically rigorous results on the phenomenon of **localization**. We refer to the bibliography where we chose some classics, some recent articles as well as books on the subject. However, all these results provide only one part of the picture that is accepted since the groundbreaking work [4, 79] by Anderson, Mott and Twose: one expects a metal insulator transition. This effect is supposed to depend upon the dimension and the general picture is as follows: Once translated into the language of spectral theory there is a transition from

Fig. 15.1. Metal insulator transition.

a **localized phase** that exhibits pure point spectrum (= only bound states = no transport) to a **delocalized phase** with absolutely continuous spectrum (= scattering states = transport). What has been proven so far is the occurrence of the former phase, well known under the name of localization. The missing part, delocalization, has not been settled for genuine random models.

There is need for an immediate disclaimer or, put differently, for an explanation of what I mean by "genuine".

An instance where a metal insulator transition has been verified rigorously is supplied by the **almost Mathieu operator**, a model with modest disorder for which the parameter that triggers the transition is the strength of the coupling. As references let us mention [6, 40, 57, 58, 59, 73] where the reader can find more about the literature on this true evergreen. Quite recently it has attracted a lot of interest especially among harmonic analysts; see [7, 8, 9, 10, 11, 12, 13, 14, 15, 44, 82]

The underlying Hilbert space is $l^2(\mathbb{Z})$. Consider parameters $\alpha, \lambda, \theta \in \mathbb{R}$ and define the self-adjoint, bounded operator $h_{\alpha,\lambda,\theta}$ by

$$(h_{\alpha,\lambda,\theta}u)(n) = u(n+1) + u(n-1) + \lambda\cos(2\pi(\alpha n + \theta))u(n),$$

for $u = (u(n))_{n\in\mathbb{Z}} \in l^2(\mathbb{Z})$.

Note that this operator is a discrete Schrödinger operator with a potential term with the coupling constant λ in front and the discrete analog of the Laplacian. For irrational α the potential term is an almost periodic function on $\mathbb{Z}$.

Basically, there is a metal insulator transition at the critical value 2 for the coupling constant λ. Since these operators are very close to being periodic, one can fairly label them as poorly disordered. Moreover, the proof of delocalization boils down to the proof of localization for a "dual operator" that happens to have the same form. In this sense, the almost Mathieu operator is not a genuine random model.

A second instance, where a delocalized phase is proven to exist is the Bethe lattice. See Klein's paper [61].

Quite recently, an order parameter has been introduced by Germinet and Klein to characterize the range of energies where a multiscale scenario provides a proof of a localized regime, [42]. In their work the important parameter is the energy.

However, as we already pointed out above, for genuine random models, there is no rigorous proof of the existence of a transition or even of the appearance of spectral components other than pure point, so far. This is a quite strange situation: the unperturbed problem exhibits extended states and purely a.c. spectrum but for the perturbed problem one can prove the opposite spectral behavior only.

In this survey we are dealing with models that are not transitive in the sense that the influence of the random potential is not uniform in space. The precise meaning of this admittedly vague description differs from case to case but will be clear for each of them.

15.2 Sparse random potentials

The term sparse potentials is mostly known for potentials that have been introduced in the 1970s by Pearson [81] to construct Schrödinger operators on the line with singular continuous spectrum. To use similar geometries to obtain a metal insulator transition can be traced back to Molchanov, Molchanov and Vainberg [77, 78] and Krishna [70, 71], see also [63, 64, 72]. We have been strongly influenced by the paper [49] from which we take the random operator in $L^2(\mathbb{R}^d)$, **Model I:**

$$H(\omega) = -\Delta + V_\omega, \text{ where } V_\omega(x) = \sum_{k \in \mathbb{Z}^m} \xi_k(\omega) f(x - k),$$

$f \leq 0$ is a compactly supported single site potential and the ξ_k are independent Bernoulli variables with $p_k := \mathbb{P}\{\xi_k = 1\}$.

To understand the appearance of a metallic regime, we recall the following facts from scattering theory:

We write $-\Delta = H_0$ so that the operators we are interested in can be written as $H = H_0 + V$. By $\sigma_{ac}(H)$ we denote the absolutely continuous spectrum, related to delocalized states.

Theorem 1 (Cooks criterion) *If for some $T_0 > 0$ and all ϕ in a dense set*

$$\int_{T_0}^{\infty} \|V e^{-itH_0}\phi\| dt < \infty, \qquad (*)$$

then $\Omega_- := \lim_{t \to \infty} e^{itH} e^{-itH_0}$ exists and, consequently, $[0, \infty) \subset \sigma_{ac}(H)$, i.e., there are scattering states for H and any nonnegative energy.

The typical application rests on the fact that $(*)$ is satisfied if

$$|V(x)| \leq C(1 + |x|)^{-(1+\epsilon)}, \qquad (**)$$

a condition that obviously fails to hold for almost every V_ω provided the p_k are not summable. However, the following nice result holds; see Hundertmark and Kirsch [49] who also provided the absolutely correct name:

Theorem 2 (Almost surely free lunch theorem) *Assume that*

$$W(x) := \left(\mathbb{E}(V_\omega(x)^2)\right)^{\frac{1}{2}} \overset{!}{\leq} C(1 + |x|)^{-(1+\epsilon)}.$$

Then V_ω satisfies Cook's criterion for a.e. ω.

The proof is so elegant and short that we can not resist to reproduce it here.

Proof.

$$\mathbb{E}\left(\int_{T_0}^{\infty} \|V_\omega e^{-itH_0}\phi\| dt\right)$$

$$= \int_{T_0}^{\infty} \mathbb{E}\left(\int V_\omega(x)^2 |e^{-itH_0}\phi(x)|^2 dx\right)^{\frac{1}{2}} dt$$

$$= \int_{T_0}^{\infty} \left(\mathbb{E}\int V_\omega(x)^2 |e^{-itH_0}\phi(x)|^2 dx\right)^{\frac{1}{2}} dt$$

$$\leq \int_{T_0}^{\infty} \left(\int \mathbb{E}(V_\omega(x)^2) |e^{-itH_0}\phi(x)|^2 dx]\right)^{\frac{1}{2}} dt$$

$$= \int_{T_0}^{\infty} \|W(x) e^{-itH_0}\phi\| dt \qquad \qquad \square$$

One can apply this result if the p_k decay fast enough to guarantee sufficient decay of $W(x)$. On the other hand one wants to have that $\sum_k p_k = \infty$, since otherwise V_ω has compact support a.s. by the Borel–Cantelli Lemma.

For fixed $d \geq 3$ and $\frac{d}{2} + \frac{1}{2} < \alpha < d$ and $p_k \sim k^{-\alpha}$ one can moreover control the essential spectrum below 0 as done in [49]: the negative essential spectrum consists of a sequence of energies that can at most accumulate at 0. Therefore, the negative spectrum is pure point. This can be summarized in the following picture:

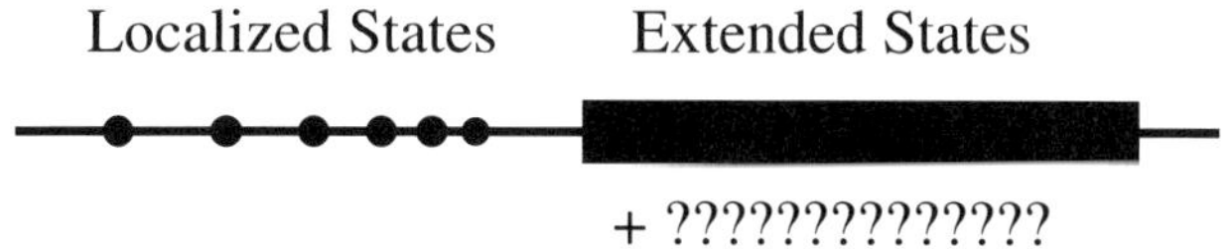

Fig. 15.2. The spectral picture for the sparse model I

We refer the reader to [49] for more on sparse random potentials, especially for models for which the negative spectrum has a richer structure and contains intervals.

Remarks 3 *In [17] we prove absence of an (absolutely) continuous spectrum outside the spectrum of the unperturbed operator for certain random sparse models reminiscent of Model I above and Model II from [49] but considerably more general. We use the techniques from [52, 90, 91].*

15.3 Random surface models

Consider the following self-adjoint random operator in $L^2(\mathbb{R}^d)$ or $\ell^2(\mathbb{Z}^d)$, $\mathbb{R}^d = \mathbb{R}^m \times \mathbb{R}^{d-m}$:

$$H(\omega) = -\Delta + V_\omega, \text{ where } V_\omega(x) = \sum_{k \in \mathbb{Z}^m} q_k(\omega)f(x - (k,0)),$$

the q_k are i.i.d. random variables and $f \geq 0$ is a single site potential that satisfies certain technical assumptions. This leads to the following geometry characterizing random surface models. Sometimes the upper half-plane is considered only.

There is a lot of literature, mostly on the discrete case, using a decomposition into a bulk and a surface term see; [5, 18, 21, 46, 50, 51, 54, 53, 55, 56].

The moral of the story is the appearance of a metal insulator transition at the edges of the unperturbed operator. We now concentrate on the continuum case, where we only know of [16, 49] as references. The existence of an a.c. component is proven in [49]. In the following, we present the result from

[16], giving strong dynamical localization. Similar but somewhat different results have been announced in [49]. As discussed there, an additional Dirichlet boundary condition "stabilizes" the spectrum so that the appearance of a negative spectrum requires a certain strength of the random perturbation. Therefore, proving localization at negative energies is easier (compared to the case without Dirichlet boundary conditions) since one is automatically dealing with a "large coupling" regime.

In [16], no use is made of an additional Dirichlet b.c. and we have the following picture:

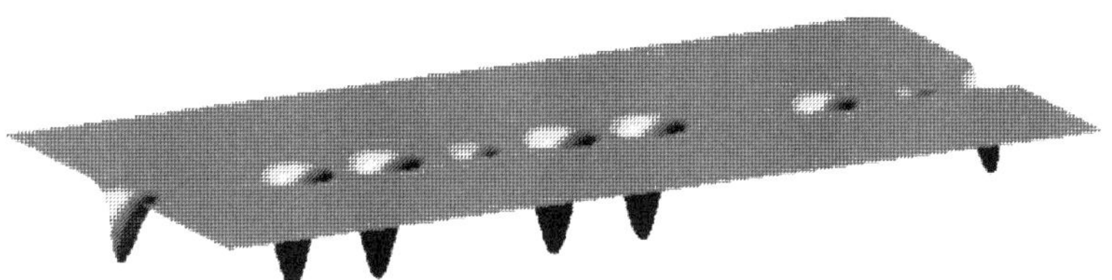

Fig. 15.3. A typical realization of a continuum random surface potential.

It is not hard to see that

$$\sigma(H(\omega)) = [E_0, \infty) \text{ where } E_0 = \inf \sigma(-\Delta + q_{\min} \cdot f^{\mathrm{per}}),$$

and

$$f^{\mathrm{per}} = \sum_{k \in \mathbb{Z}^m} f(x - (k, 0))$$

denotes the periodic continuation of f along the surface. Near the bottom of the spectrum E_0 one expects localization, i.e., suppression of transport as is typical for insulators. For nonnegative energies one expects extended states. To stress the existence of a metallic phase let us cite Theorem 4.3 of [49] t

Theorem 4 *Let $H(\omega)$ be as below. Then we have, for every $\omega \in \Omega$: $[0, \infty) \subset \sigma_{\mathrm{ac}}(H(\omega))$.*

The idea of the proof is that a wave packet with velocity pointing away from the surface will escape the influence of the surface potential and is asymptotically free. The rigorous implementation of this idea uses Enss' technique from scattering theory.

The model   *(1) $0 < m < d$ and points in $\mathbb{R}^d = \mathbb{R}^m \times \mathbb{R}^{d-m}$ are written as pairs, if convenient;*
(2) The single site potential $f \geq 0$, $f \in L^p(\mathbb{R}^d)$ where $p \geq 2$ if $d \leq 3$ and $p > d/2$ if $d > 3$, and $f \geq \sigma > 0$ on some open set $U \neq \emptyset$ for some $\sigma > 0$.

(3) The q_k are i.i.d. random variables distributed with respect to a probability measure μ on $\mathbb{R}$, such that $\operatorname{supp}\mu = [q_{min}, 0]$ with $q_{min} < 0$.

We will sometimes need further assumptions on the single site distribution μ:

(4) μ is Hölder continuous, i.e., there are constants $C, \alpha > 0$ such that

$$\mu[a, b] \le C(b - a)^\alpha \text{ for } q_{min} \le a \le b \le 0.$$

(5) Disorder assumption: there exist $C, \tau > 0$ such that

$$\mu[q_{min}, q_{min} + \varepsilon] \le C \cdot \varepsilon^\tau \text{ for } \varepsilon > 0.$$

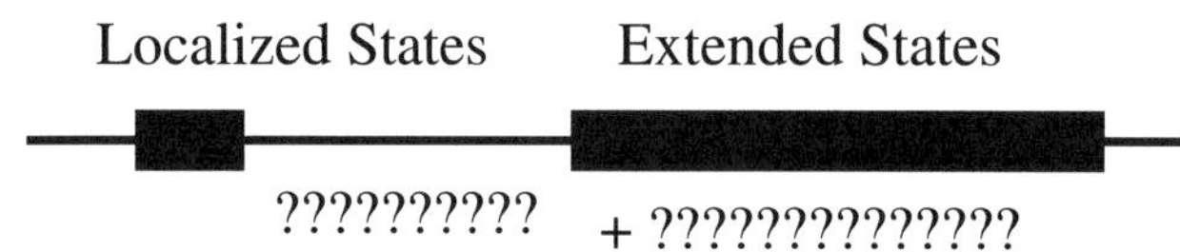

Fig. 15.4. Conclusion and open problems for the continuum surface model.

What follows is the main result of [16].

Theorem 5 *Let $H(\omega)$ be as above with $\tau > d/2$ and assume that $E_0 < 0$.*

(a) There exists an $\varepsilon > 0$ such that in $[E_0, E_0 + \varepsilon]$ the spectrum of $H(\omega)$ is pure point for almost every $\omega \in \Omega$, with exponentially decaying eigenfunctions.

(b) Assume that $p < 2(2\tau - m)$. Then there exists an $\varepsilon > 0$ such that in $[E_0, E_0 + \varepsilon] = I$ we have strong dynamical localization in the sense that for every compact set $K \subset \mathbb{R}^d$:

$$\mathbb{E}\{\sup_{t>0} \|\,|X|^p e^{-itH(\omega)} P_I(H(\omega))\chi_K\|\} < \infty.$$

A consequence is a pure point spectrum in the interval $[E_0, E_0 + \varepsilon] = I$. Together with the previous result on extended states we get the picture from Figure 15.4 that still leaves open some important questions.

Remarks 6 *(1) That we have to assume $E_0 < 0$ was pointed out to us by J. Voigt. Since we allow arbitrary m and $d - m$, a negative perturbation will not automatically create a negative spectrum.*

(2) In [17] we will present results that cover the negative spectrum for the model above using techniques from [52, 90, 91]. So far this works only for $m = 1$ but the proofs allow quite arbitrary background perturbations.

References

1. M. Aizenman and S.A. Molchanov, Localization at large disorder and at extreme energies: An elementary derivation, *Commun. Math. Phys.*, **157** (1993), 245–278.
2. M. Aizenman and G.M. Graf, Localization bounds for an electron gas, *J. Phys. A: Math. Gen.* **31** (1998), 6783–6806.
3. M. Aizenman, A. Elgart, S. Naboko, J.H. Schenker, G. Stolz, Moment analysis for localization in random Schroedinger operators, `eprint`, `arXiv`, `math-ph/0308023`
4. P.W. Anderson, Absence of diffusion in certain random lattices, *Phys. Rev.* **109** (1958), 1492–1505.
5. Bentosela, F., Briet, Ph., Pastur, L., On the spectral and wave propagation properties of the surface Maryland model. *J. Math. Phys.* **44**:1 (2003), 1–35.
6. J. Bourgain, M. Goldstein, On nonperturbative localization with quasiperiodic potential, *Ann. of Math.*, **152** (2000), 835–879.
7. J. Bourgain, On the spectrum of lattice Schrödinger operators with deterministic potential. II. Dedicated to the memory of Tom Wolff, *J. Anal. Math.* **88** (2002), 221–254.
8. J. Bourgain, M. Goldstein and W. Schlag, Anderson localization for Schrödinger operators on $\mathbf{Z}^2$ with quasi-periodic potential, *Acta Math.* **188**:1 (2002), 41–86.
9. J. Bourgain, New results on the spectrum of lattice Schrödinger operators and applications. In *Mathematical results in quantum mechanics (Taxco, 2001)*, Contemp. Math., Vol. 307, Amer. Math. Soc., Providence, RI, 2002, 27–38.
10. J. Bourgain, On the spectrum of lattice Schrödinger operators with deterministic potential. Dedicated to the memory of Thomas H. Wolff, *J. Anal. Math.* **87** (2002), 37–75.
11. J. Bourgain, Exposants de Lyapounov pour opérateurs de Schrödinger discrètes quasi-périodiques. *C. R. Math. Acad. Sci. Paris* **335**:6 (2002), 529–531.
12. J. Bourgain and S. Jitomirskaya, Continuity of the Lyapunov exponent for quasiperiodic operators with analytic potential. Dedicated to David Ruelle and Yasha Sinai on the occasion of their 65th birthdays. *J. Statist. Phys.* **108**:5,6 (2002), 1203–1218.
13. J. Bourgain and S. Jitomirskaya, Absolutely continuous spectrum for 1D quasiperiodic operators. *Invent. Math.* **148**:3 (2002), 453–463.
14. J. Bourgain, On random Schrödinger operators on $\mathbb{Z}^2$. *Discrete Contin. Dyn. Syst.* **8**:1 (2002), 1–15.
15. J. Bourgain, M. Goldstein and W. Schlag, Anderson localization for Schrödinger operators on $\mathbb{Z}$ with potentials given by the skew-shift. *Comm. Math. Phys.* **220**:3 (2001), 583–621.
16. A. Boutet de Monvel, P. Stollmann, Dynamical localization for continuum random surface models, *Arch. Math.*, **80** (2003), 87–97.
17. A. Boutet de Monvel, P. Stollmann and G. Stolz, Absence of continuous spectral types for certain nonstationary random models. Preprint, 2004.
18. A. Boutet de Monvel, A. Surkova, Localisation des états de surface pour une classe d'opérateurs de Schrödinger discrets à potentiels de surface quasi-périodiques, *Helv. Phys. Acta* **71**:5 (1998), 459–490.
19. R. Carmona and J. Lacroix, *Spectral Theory of Random Schrödinger Operators*, Birkhäuser, Boston, 1990.

20. R. Carmona and A. Klein and F. Martinelli, Anderson localization for Bernoulli and other singular potentials, *Commun. Math. Phys.*, **108** (1987), 41–66.

21. A. Chahrour and J. Sahbani, On the spectral and scattering theory of the Schrödinger operator with surface potential, *Rev. Math. Phys.* **12**:4 (2000), 561–573.

22. J.M. Combes and P.D. Hislop, Localization for some continuous, random Hamiltonians in d-dimensions, *J. Funct. Anal.*, **124** (1994), 149–180.

23. J.M. Combes, and P.D. Hislop, Landau Hamiltonians with random potentials: Localization and density of states, *Commun. Math. Phys.*, **177** (1996), 603–630.

24. D. Damanik and P. Stollmann, Multi-scale analysis implies strong dynamical localization. *GAFA, Geom. Funct. Anal.* **11** (2001), 11–29.

25. R. del Rio, S. Jitomirskaya, Y. Last and B. Simon, What is localization? *Phys. Rev. Letters* **75** (1995), 117–119.

26. R. del Rio, S. Jitomirskaya, Y. Last and B. Simon, Operators with singular continuous spectrum IV: Hausdorff dimensions, rank one perturbations and localization, *J. d'Analyse Math.* **69** (1996), 153–200.

27. F. Delyon, Y. Levy and B. Souillard, Anderson localization for multidimensional systems at large disorder or low energy, *Commun. Math. Phys.* **100** (1985), 463–470.

28. F. Delyon, Y. Levy and B. Souillard, Approach à la Borland to multidimensional localisation, *Phys. Rev. Lett.* **55** (1985), 618–621.

29. T.C. Dorlas, N. Macris and J.V. Pulé, Localization in a single-band approximation to random Schrödinger operators in a magnetic field, *Helv. Phys. Acta*, **68** (1995), 329–364.

30. H. von Dreifus, On the effects of randomness in ferromagnetic models and Schrödinger operators. NYU, Ph.D. Thesis (1987).

31. H. von Dreifus and A. Klein, A new proof of localization in the Anderson tight binding model, *Commun. Math. Phys.* **124** (1989), 285–299.

32. W.G. Faris, Random waves and localization. *Notices Amer. Math. Soc.* **42**:8 (1995), 848–853.

33. A. Figotin and A. Klein, Localization of classical waves I: Acoustic waves, *Commun. Math. Phys.* **180** (1996), 439–482.

34. W. Fischer, H. Leschke and P. Müller, Spectral localization by Gaussian random potentials in multi-dimensional continuous space, *J. Stat. Phys.* **101** (2000), 935–985.

35. J. Fröhlich, Les Houches lecture, in *Critical Phenomena, Random Systems, Gauge Theories*, K. Osterwalder and R. Stora, eds. North-Holland, 1984.

36. J. Fröhlich, F. Martinelli, E. Scoppola and T. Spencer, Constructive Proof of Localization in the Anderson Tight Binding Model, *Commun. Math. Phys.* **101** (1985), 21–46.

37. J. Fröhlich and T. Spencer, Absence of Diffusion in the Anderson Tight Binding Model for Large Disorder or Low Energy, *Commun. Math. Phys.* **88** (1983), 151–184.

38. F. Germinet, Dynamical Localization II with an Application to the Almost Mathieu Operator. *J. Statist. Phys.* **98**:5,6 (2000), 1135–1148.

39. F. Germinet and S. de Bievre, Dynamical Localization for Discrete and Continuous Random Schrödinger operators, *Commun. Math. Phys.* **194** (1998), 323–341.

40. F. Germinet and S. Jitomirskaya, Strong dynamical localization for the almost Mathieu model. *Rev. Math. Phys.* **13**:6 (2001), 755–765.

41. F. Germinet and A. Klein, Bootstrap multiscale analysis and localization in random media. *Comm. Math. Phys.* **222**:2 (2001), 415–448.
42. F. Germinet and H. Klein, A characterization of the Anderson metal-insulator transport transition; submitted.
43. I. Ya. Goldsheidt, S.A. Molchanov and L. A. Pastur, Typical one-dimensional Schrödinger operator has pure point spectrum, *Funktsional. Anal. i Prilozhen.* **11**:1 (1977), 1–10 ; Engl. transl. in *Functional Anal. Appl.* **11** (1977).
44. M. Goldstein, W. Schlag, Hölder continuity of the integrated density of states for quasi-periodic Schrödinger equations and averages of shifts of subharmonic functions, *Ann. of Math. (2)* **154**:1 (2001), 155–203.
45. G.M. Graf, Anderson localization and the space-time characteristic of continuum states, *J. Stat. Phys.* **75** (1994), 337–343.
46. V. Grinshpun, Localization for random potentials supported on a subspace, *Lett. Math. Phys.* **34**:2 (1995), 103–117.
47. H. Holden and F. Martinelli, On the absence of diffusion for a Schrödinger operator on $L^2(R^\nu)$ with a random potential, *Commun. Math. Phys.* **93** (1984), 197–217.
48. D. Hundertmark, On the time-dependent approach to Anderson localization, *Math. Nachrichten,* **214** (2000), 25–38.
49. D. Hundertmark, W. Kirsch, Spectral theory of sparse potentials, in *Stochastic Processes, Physics and Geometry: New Interplays, I (Leipzig, 1999)*, Amer. Math. Soc., Providence, RI, 2000, 213–238.
50. V. Jakšić, Y. Last, Corrugated surfaces and a.c. spectrum, *Rev. Math. Phys.* **12**(11) (2000), 1465–1503.
51. V. Jakšić, Y. Last, Spectral structure of Anderson type hamiltonians, *Inv. Math.* **141**:3 (2000), 561–577
52. I. McGillivray, P. Stollmann and G. Stolz, Absence of absolutely continuous spectra for multidimensional Schrödinger operators with high barriers, *Bull. London Math. Soc.* **27**:2 (1995), 162–168.
53. V. Jakšić, S. Molchanov: On the surface spectrum in dimension two, *Helv. Phys. Acta* **71**:6 (1998), 629–657.
54. V. Jakšić, S. Molchanov, On the spectrum of the surface Maryland model, *Lett. Math. Phys.* **45**:3 (1998), 189–193.
55. V. Jakšić, S. Molchanov, Localization of surface spectra, *Comm. Math. Phys.* **208**:1 (1999), 153–172.
56. V. Jakšić, S. Molchanov, L. Pastur, On the propagation properties of surface waves, in *Wave Propagation in Complex Media (Minneapolis, MN, 1994)*, IMA Math. Appl., Vol. 96, Springer, New York, 1998, 143–154.
57. S. Jitomirskaya, Almost everything about the almost Mathieu operator II. In: *Proc. of the XIth International Congress of Mathematical Physics*, Paris 1994, Int. Press, 1995, 366–372.
58. S. Jitomirskaya and Y. Last, Anderson localization for the almost Mathieu operator III. Semi-uniform localization, continuity of gaps and measure of the spectrum, *Commun. Math. Phys.* **195** (1998), 1–14.
59. S. Jitomirskaya, Metal-Insulator transition for the Almost Mathieu operator, *Annals of Math.,* **150** (1999), 1159–1175.
60. S. John, The localization of light and other classical waves in disordered media, *Comm. Cond. Mat. Phys.* **14**:4 (1988), 193–230.
61. A. Klein: Extended states in the Anderson model on the Bethe lattice. *Adv. Math.* **133**:1 (1998), 163–184.

62. W. Kirsch, Wegner estimates and Anderson localization for alloy-type potentials, *Math. Z.* **221** (1996), 507–512.

63. W. Kirsch, Scattering theory for sparse random potentials. *Random Oper. Stochastic Equations* **10**:4 (2002), 329–334.

64. W. Kirsch, M. Krishna and J. Obermeit, Anderson model with decaying randomness: Mobility edge, *Math. Z.*, **235** (2000), 421–433.

65. W. Kirsch, P. Stollmann, G. Stolz: Localization for random perturbations of periodic Schrödinger operators, *Random Oper. Stochastic Equations* **6**:3 (1998), 241–268.

66. W. Kirsch, P. Stollmann, G. Stolz: Anderson localization for random Schrödinger operators with long range Interactions, *Comm. Math. Phys.* **195**:3 (1998), 495–507.

67. F. Kleespies and P. Stollmann, Localization and Lifshitz tails for random quantum waveguides, *Rev. Math. Phys.* **12**:10 (2000), 1345–1365.

68. F. Klopp, Localization for semiclassical Schrödinger operators II: The random displacement model, *Helv. Phys. Acta* **66** (1993), 810–841.

69. F. Klopp, Localization for some continuous random Schrödinger operators, *Commun. Math. Phys.* **167** (1995), 553–569.

70. M. Krishna: Anderson model with decaying randomness: Existence of extended states, *Proc. Indian Acad. Sci. Math. Sci.* **100**:3, 285–294.

71. M. Krishna: Absolutely continuous spectrum for sparse potentials, *Proc. Indian Acad. Sci. Math. Sci.* **103**:3 (1993), 333–339.

72. M. Krishna, K.B. Sinha, Spectra of Anderson type models with decaying randomness. *Proc. Indian Acad. Sci. Math. Sci.* **111**:2 (2001), 179–201.

73. Y. Last, Almost everything about the almost Mathieu operator I, in *Proc. of the XIth International Congress of Mathematical Physics,* Paris 1994, Int. Press, 1995, 373–382.

74. F. Martinelli and H. Holden, On absence of diffusion near the bottom of the spectrum for a random Schrödinger operator on $L^2(\mathbb{R}^\nu)$, *Commun. Math. Phys.* **93** (1984), 197–217.

75. F. Martinelli and E. Scoppola, Introduction to the mathematical theory of Anderson localization, *Rivista Nuovo Cimento* **10** (1987), 10.

76. F. Martinelli and E. Scoppola, Remark on the absence of the absolutely continuous spectrum for d-dimensional Schrödinger operator with random potential for large disorder or low energy, *Commun. Math. Phys.* **97** (1985), 465–471; Erratum ibid. **98** (1985), 579.

77. S. Molchanov, Multiscattering on sparse bumps, in *Advances in Differential Equations and Mathematical Physics (Atlanta, GA, 1997)*, Contemp. Math., Vol. 217, Amer. Math. Soc., Providence, RI, 1998, 157–181.

78. S. Molchanov, B. Vainberg, Multiscattering by sparse scatterers, in *Mathematical and Numerical Aspects of Wave Propagation (Santiago de Compostela, 2000)*, SIAM, Philadelphia, PA, 2000, 518–522.

79. N.F. Mott and W.D. Twose, The theory of impurity conduction, *Adv. Phys.* **10** (1961), 107–163.

80. L.A. Pastur and A. Figotin, *Spectra of Random and Almost-Periodic Operators*, Springer-Verlag, Berlin, 1992.

81. D. Pearson, Singular continuous measures in scattering theory, *Commun. Math. Phys.* **60** (1976), 13–36.

82. W. Schlag, C. Shubin, T. Wolff, Frequency concentration and location lengths for the Anderson model at small disorders. Dedicated to the memory of Tom Wolff, *J. Anal. Math.* **88** (2002), 173–220.

83. C. Shubin, R. Vakilian and T. Wolff, Some harmonic analysis questions suggested by Anderson–Bernoulli models, *GAFA* **8** (1998), 932–964.

84. B. Simon and T. Wolff, Singular continuous spectrum under rank one perturbations and localization for random Hamiltonians, *Comm. Pure Appl. Math.* **39** (1986), 75–90.

85. R. Sims and G. Stolz, Localization in one dimensional random media: a scattering theoretic approach, *Comm. Math. Phys.* **213** (2000), 575–597.

86. T. Spencer, The Schrödinger equation with a random potential – A mathematical review, in *Critical Phenomena, Random Systems, Gauge Theories*, K. Osterwalder and R. Stora, eds. North Holland, 1984.

87. T. Spencer, Localization for random and quasi-periodic potentials, *J. Stat. Phys.* **51** (1988), 1009.

88. P. Stollmann, *Caught by Disorder – Bound States in Random Media*, Progress in Math. Phys., Vol. 20, Birkhäuser, Boston, 2001.

89. P. Stollmann, Wegner estimates and localization for continuum Anderson models with some singular distributions, *Arch. Math.* **75**:4 (2000), 307–311.

90. P. Stollmann and G. Stolz, Singular spectrum for multidimensional operators with potential barriers. *J. Operator Theory* **32** (1994), 91–109.

91. G. Stolz, Localization for Schrödinger operators with effective barriers. *J. Funct. Anal.* **146**:2 (1997), 416–429.

92. G. Stolz, Localization for random Schrödinger operators with Poisson potential, *Ann. Inst. Henri Poincaré* **63**, 297–314 (1995).

93. D.J. Thouless, Introduction to disordered systems. *Phènoménes Critiques, Systèmes Aléatoires, Théories de Jauge, Part I, II (Les Houches, 1984)*, North-Holland, Amsterdam, 1986, 681–722.

94. I. Veselic, Localisation for random perturbations of periodic Schrödinger operators with regular Floquet eigenvalues. *Ann. Henri Poincare* **3**:2 (2002), 389–409.

95. W.M. Wang, Microlocalization, percolation, and Anderson localization for the magnetic Schrödinger operator with a random potential, *J. Funct. Anal.* **146** (1997), 1–26.

96. H. Zenk, Anderson localization for a multidimensional model including long range potentials and displacements. *Rev. Math. Phys.* **14**:2 (2002), 273–302.

On a Rigorous Proof of the Joos–Zeh Formula for Decoherence in a Two-body Problem

A. Teta

Dipartimento di Matematica Pura e Applicata
Università di l'Aquila
I-67100 l'Aquila, Italy

Summary. We consider a simple one-dimensional system consisting of two particles interacting with a δ-potential and we discuss a rigorous derivation of the asymptotic wave function of the system in the limit of small mass ratio. We apply the result for the explicit computation of the decoherence effect induced on the heavy particle in a concrete example of quantum evolution.

In this note we shall briefly discuss the asymptotic form of the wave function of a two-particle system in the limit of small mass ratio.

The interest in this problem is motivated by the analysis initiated by Joos and Zeh ([JZ]) of the mechanism of decoherence induced on a heavy particle by the scattering of light particles (see also [GF],[T],[BGJKS],[GJKKSZ],[D]).

The basic idea for the analysis of the process is that the small mass ratio produces a separation of two characteristic time scales, one slow for the dynamics of the heavy particle and the other fast for the light ones.

Following this line, in [JZ] the elementary scattering event between a light and a heavy particle is described by the instantaneous transition

$$\phi(R)\chi(r) \to \phi(R)(S_R\chi)(r) \tag{16.1}$$

where ϕ and χ are the initial wave functions of the heavy and the light particle respectively and S_R is the scattering operator of the light particle when the heavy particle is considered fixed in its initial position R.

In (16.1) the initial state is chosen in the form of a product state, i.e., no correlation is assumed at time zero; moreover the final state is computed in a zero-th order adiabatic approximation for small values of the mass ratio.

Formula (16.1) gives a simple and clear description of the scattering event; nevertheless the approximation chosen is rather crude in the sense that time zero of the heavy particle corresponds to infinite time of the light one and the evolution in time of the system is completely neglected.

In order to restore the time evolution, in [JZ] the formula is modified introducing by hand the internal dynamics of the heavy particle.

198   A. Teta

Our aim in this note is to discuss how the complete Joos and Zeh formula, i.e., modified taking into account the internal motion of the heavy particle, can be rigorously derived from the Schrödinger equation of the whole two-particle system.

In particular, using the result proved in [DFT], we shall write the asymptotic form of the wave function of the system approximating the true wave function in a specific scaling limit (involving the mass ratio and the strength of the interaction) with an explicit control of the error.

Furthermore we shall apply the result for the analysis of decoherence in a concrete example of quantum evolution.

More precisely, we shall consider an initial state with the heavy particle in a coherent superposition of two spatially separated wave packets with opposite momentum and the light one localized far on the left with a positive momentum.

Under precise assumptions on the relevant physical parameters (spreading and momentum of the light particle, effective range of the interaction, spreading and separation of the wave packets of the heavy particle) we shall derive an approximated form of the wave function of the system, describing the typical entangled state with the wave of the light particle split into a reflected and a transmitted part by each wave packet of the heavy particle.

As a consequence, the reduced density matrix for the heavy particle will show unperturbed diagonal terms and off-diagonal terms reduced by a factor Λ less than 1, which gives a measure of the decoherence effect induced on the heavy particle.

Due to a phase shift, the waves reflected by the two wave packets are shown to be approximately orthogonal if the separation of the two wave packets is sufficiently large and then the factor Λ can be explicitly computed in terms of the transmission probability for the light particle subject to a δ-potential placed at a fixed position.

We want to emphasize that, at each step, the approximate formulas are obtained through rather elementary estimates with explicit control of the error. In this sense the model may be considered of pedagogical relevance for the analysis of the mechanism of decoherence.

Let us introduce the model. The hamiltonian of the two-particle system is given by

$$H = -\frac{\hbar^2}{2M}\Delta_R - \frac{\hbar^2}{2m}\Delta_r + \alpha_0\delta(r - R) \tag{16.2}$$

where M and m denote the masses of the heavy and the light particle respectively and $\alpha_0 > 0$ is the strength of the interaction.

The hamiltonian (16.2) is a well-defined self-adjoint and positive operator in $L^2(\mathbb{R}^2)$ (see e.g., [AGH-KH]) and moreover it is a solvable model, in the sense that the corresponding generalized eigenfunctions and the unitary group can be explicitly computed ([S]).

We consider an initial state given in a product form

$$\psi_0(r, R) = \phi(R)\chi(r), \qquad \phi, \chi \in \mathcal{S}(\mathbb{R}) \tag{16.3}$$

where $\mathcal{S}(\mathbb{R})$ is the Schwartz space, and we denote by $\epsilon = \frac{m}{M}$ the mass ratio which is the small parameter of the model.

We are interested in the asymptotic form of the solution of the corresponding Schrödinger equation

$$i\hbar \frac{\partial \psi_t^\epsilon}{\partial t} = H\psi_t^\epsilon \tag{16.4}$$

when $\epsilon \to 0$ and $\hbar$, M and the parameter

$$\alpha = \frac{m\alpha_0}{\hbar^2} \tag{16.5}$$

are kept fixed. Notice that α^{-1} is a length with the physical meaning of an effective range of the interaction.

Rescaling the time according to

$$\tau = \frac{\hbar}{M} t \tag{16.6}$$

the Schrödinger equation can be more conveniently written as

$$i\frac{\partial \psi_\tau^\epsilon}{\partial \tau} = -\frac{1}{2}\Delta_R\, \psi_\tau^\epsilon + \frac{1}{\epsilon}\left(-\frac{1}{2}\Delta_r + \alpha\delta(r - R)\right)\psi_\tau^\epsilon \tag{16.7}$$

From (16.7) it is clear that for $\epsilon \to 0$ the kinetic energy of the heavy particle can be considered as a small perturbation.

The situation is similar to the Born–Oppenheimer approximation, with the relevant difference that here the light particle is not in a bound state and so it cannot produce any effective potential for the heavy particle.

Then, for $\epsilon \to 0$, one should expect a scattering regime for the light particle in presence of the heavy one in a fixed position and a free motion for the heavy particle.

In fact, the asymptotics for $\epsilon \to 0$ of ψ_τ^ϵ is characterized in the following proposition ([DFT]).

Theorem 1. *For any initial state (16.3) and any $\tau > 0$ one has*

$$\|\psi_\tau^\epsilon - \psi_\tau^a\| < \left(\frac{A}{\tau} + B\right)\epsilon \tag{16.8}$$

where

$$\psi_\tau^a(r, R) = \frac{1}{\sqrt{i\epsilon^{-1}\tau}} e^{i\frac{\epsilon}{2\tau}r^2} \int dx e^{-i\tau H_0}(R - x)\phi(x) \left[(\Omega_+^x)^{-1}\chi\right]\tilde{}\left(\frac{r}{\epsilon^{-1}\tau}\right) \tag{16.9}$$

and $H_0 = -\frac{1}{2}\Delta$, the symbol $\tilde{}$ denotes the Fourier transform, Ω_+^x is the wave operator associated to the one-particle hamiltonian $H_x = -\frac{1}{2}\Delta + \alpha\delta(\cdot - x)$, for any $x \in \mathbb{R}$; moreover A, B are positive, time-independent constants whose dependence on the strength of the interaction and on the initial state are explicitly given.

We want to apply the result stated in Theorem 1 to the analysis of decoherence in a concrete example of quantum evolution. In particular we consider an initial state $\psi_0(R, r) = \phi(R)\chi(r)$ of the form

$$\phi(R) = \frac{1}{\sqrt{2}}\left(f_\sigma^+(R) + f_\sigma^-(R)\right), \tag{16.10}$$

$$f_\sigma^\pm(R) = \frac{1}{\sqrt{\sigma}}f\left(\frac{R \pm R_0}{\sigma}\right)e^{\pm iP_0 R}, \quad \sigma, R_0, P_0 > 0, \qquad R_0 > 2\sigma, \tag{16.11}$$

$$\chi(r) = g_\delta(r) = \frac{1}{\sqrt{\delta}}g\left(\frac{r - r_0}{\delta}\right)e^{iq_0 r}, \quad \delta, q_0 > 0, r_0 < -R_0 - \sigma - \delta, \tag{16.12}$$

$$f, g \in C_0^\infty(-1, 1), \quad \|f\| = \|g\| = 1. \tag{16.13}$$

From (16.10),(16.11) one sees that the heavy particle is in a superposition state of two spatially separated wave packets, one localized in $R = -R_0$ with mean value of the momentum P_0 and the other localized in $R = R_0$ with mean value of the momentum $-P_0$.

From (16.12), the light particle is assumed localized around r_0, on the left of the wave packet f_σ^+, with positive mean momentum q_0.

Moreover, to simplify the computation, f_σ^+, f_σ^-, g_δ are chosen compactly supported and with disjoint supports.

We expect that in the time evolution of the above initial state, for $\epsilon \to 0$, the light particle will be partly reflected and partly transmitted by the two wave packets f_σ^+, f_σ^-, which approximately act as fixed scattering centers.

We shall make precise this statement introducing suitable assumptions on the physical parameters of the system.

In order to formulate the result, we introduce the reflection and transmission coefficient associated to the one-particle hamiltonian H_x (see e.g., [AGH-KH])

$$\mathcal{R}_\alpha(k) = -\frac{\alpha}{\alpha - i|k|}, \qquad \mathcal{T}_\alpha(k) = -\frac{i|k|}{\alpha - i|k|} \tag{16.14}$$

and define the transmitted and reflected part of the wave function of the light particle

$$g_T(\epsilon^{-1}\tau, r) = \mathcal{T}_\alpha\left(\frac{|r|}{\epsilon^{-1}\tau}\right) \frac{e^{i\frac{r^2}{2\epsilon^{-1}\tau}}}{\sqrt{i\epsilon^{-1}\tau}}\, \tilde{g}_\delta\left(\frac{r}{\epsilon^{-1}\tau}\right), \tag{16.15}$$

$$g_\mathcal{R}^x(\epsilon^{-1}\tau, r) = \mathcal{R}_\alpha\left(\frac{|r|}{\epsilon^{-1}\tau}\right) e^{-2ix\frac{r}{\epsilon^{-1}\tau}} \frac{e^{i\frac{r^2}{2\epsilon^{-1}\tau}}}{\sqrt{i\epsilon^{-1}\tau}}\, \tilde{g}_\delta\left(-\frac{r}{\epsilon^{-1}\tau}\right). \tag{16.16}$$

Moreover, we shall assume

$$q_0 \gg \frac{1}{\delta}, \qquad \frac{1}{\alpha} \gg \sigma. \tag{16.17}$$

The first assumption in (16.17) simply means that the wave function of the light particle is well concentrated in momentum space around q_0 (notice that δ^{-1} is the order of magnitude of the spreading of the momentum).

The second assumption in (16.17) requires that the effective range of the interaction is much larger than the spreading in position of each wave packet of the heavy particle and this means that the interaction cannot "distinguish" two different points in the supports of f_σ^+ and f_σ^-.

Then

Proposition 2. *For the initial state (16.10), (16.11), (16.12), (16.13) and any $\tau > 0$ one has*

$$\|\psi_\tau^a - \psi_\tau^e\| < c\left[(\delta q_0)^{-n} + \sigma\alpha\right] \tag{16.18}$$

for any $n \in \mathbb{N}$, where c is a numerical constant depending on g and

$$\psi_\tau^e(r, R) = \frac{1}{\sqrt{2}} \left(e^{-i\tau H_0} f_\sigma^+\right)(R) \left[g_T(\epsilon^{-1}\tau, r) + g_\mathcal{R}^{-R_0}(\epsilon^{-1}\tau, r)\right]$$

$$+ \frac{1}{\sqrt{2}} \left(e^{-i\tau H_0} f_\sigma^-\right)(R) \left[g_T(\epsilon^{-1}\tau, r) + g_\mathcal{R}^{R_0}(\epsilon^{-1}\tau, r)\right]. \tag{16.19}$$

Proof. Exploiting the explicit expression of the generalized eigenfunctions of H_x (see e.g., [AGH-KH]), for $x \in supp\, f_\sigma^\pm$, one has

$$[(\Omega_+^x)^{-1} g_\delta]^\sim(k) = \frac{1}{\sqrt{2\pi}} \int dy\, g_\delta(y) \left(e^{-iky} + \mathcal{R}_\alpha(k) e^{-ikx} e^{i|k||x-y|}\right)$$

$$= \mathcal{T}_\alpha(k)\tilde{g}_\delta(k)\theta_+(k) + \left[\tilde{g}_\delta(k) + e^{-2ikx}\mathcal{R}_\alpha(k)\tilde{g}_\delta(-k)\right]\theta_-(k) \tag{16.20}$$

where we have used the fact that $|x - y| = x - y$ for $x \in supp\, f_\sigma^\pm$, $y \in supp\, g_\delta$, the identity $1 + \mathcal{R}_\alpha = \mathcal{T}_\alpha$ and we have denoted by θ_+, θ_- the characteristic functions of the positive and negative semi-axis respectively.

From (16.9),(16.10),(16.11),(16.12),(16.19) we can write

$$\psi_\tau^a(r,R) - \psi_\tau^e(r,R)$$

$$= \frac{1}{\sqrt{2i\epsilon^{-1}\tau}} e^{i\frac{\epsilon}{2\tau}r^2} \int dx\, e^{-i\tau H_0}(R-x)(f_\sigma^+(x) + f_\sigma^-(x))\mathcal{R}_\alpha\left(\frac{r}{\epsilon^{-1}\tau}\right)$$

$$\times \left[\tilde{g}_\delta\left(\frac{r}{\epsilon^{-1}\tau}\right)\theta_-(r) - e^{-2ix\frac{r}{\epsilon^{-1}\tau}}\tilde{g}_\delta\left(-\frac{r}{\epsilon^{-1}\tau}\right)\theta_+(r)\right]$$

$$+ \frac{1}{\sqrt{2}}\int dx\, e^{-i\tau H_0}(R-x)f_\sigma^+(x)\left(g_\mathcal{R}^x\left(\frac{r}{\epsilon^{-1}\tau}\right) - g_\mathcal{R}^{-R_0}\left(\frac{r}{\epsilon^{-1}\tau}\right)\right)$$

$$+ \frac{1}{\sqrt{2}}\int dx\, e^{-i\tau H_0}(R-x)f_\sigma^-(x)\left(g_\mathcal{R}^x\left(\frac{r}{\epsilon^{-1}\tau}\right) - g_\mathcal{R}^{R_0}\left(\frac{r}{\epsilon^{-1}\tau}\right)\right)$$

$$\equiv (\zeta_1 + \zeta_2 + \zeta_3)(r,R). \tag{16.21}$$

Noticing that $\tilde{g}_\delta(k) = e^{i(q_0-k)r_0}\sqrt{\delta}\tilde{g}(\delta k - \delta q_0)$, one has

$$\|\zeta_1\|^2 = \frac{1}{2}\int dx|f_\sigma^+(x) + f_\sigma^-(x)|^2 \int dk\,|\mathcal{R}_\alpha(k)|^2\,|\tilde{g}_\delta(k)\,\theta_-(k)$$

$$- e^{-2ixk}\tilde{g}_\delta(-k)\,\theta_+(k)|^2$$

$$\leq \int dk\left(|\tilde{g}_\delta(k)\,\theta_-(k)|^2 + |\tilde{g}_\delta(-k)\,\theta_+(k)|^2\right)$$

$$= 2\int_{\delta q_0}^\infty dz|\tilde{g}(z)|^2 \tag{16.22}$$

which is $O((\delta q_0)^{-n})$ for any $n \in \mathbb{N}$. Concerning ζ_2 we have

$$\|\zeta_2\|^2 = \frac{1}{2}\int dx|f_\sigma^+(x)|^2 \int dk|\mathcal{R}_\alpha(k)|^2|\tilde{g}_\delta(k)|^2\left|e^{-2ikx} - e^{2ikR_0}\right|^2$$

$$\leq 2\sup_{x\in(-R_0-\sigma,-R_0+\sigma)}|x + R_0|^2 \int dk k^2|\mathcal{R}_\alpha(k)|^2|\tilde{g}_\delta(k)|^2$$

$$\leq 2\sigma^2\alpha^2 \tag{16.23}$$

where we have used the inequality $k^2|\mathcal{R}_\alpha(k)|^2 \leq \alpha^2$. The estimate for ζ_3 proceeds exactly in the same way, concluding the proof. $\qquad\square$

Proposition 2 shows that, under the assumptions (16.17), the wave function (16.19) of the whole system takes the form of the typical entangled state emerging from an interaction of von Neumann type between a system and an apparatus. Here the role of the apparatus is played by the light particle which is partly transmitted and partly reflected by each wave packet of the heavy particle.

In this process the light particle keeps the information about the position of the heavy particle, which is encoded in $g_\mathcal{R}^{-R_0}$ and $g_\mathcal{R}^{R_0}$, i.e., in the reflected waves by f_σ^+ and f_σ^- respectively.

As it should be expected in a one-dimensional scattering process, such reflected waves have an identical spatial localization and they only differ for a phase factor.

In order to discuss the decoherence effect induced on the heavy particle one has to consider the density matrix of the whole system $\rho_\tau^\epsilon(r, R, r', R')$ and then one should compute the reduced density matrix for the heavy particle

$$\hat{\rho}_\tau^\epsilon(R, R') \equiv \int dr \rho_\tau^\epsilon(r, R, r, R') \equiv \int dr \psi_\tau^\epsilon(r, R)\overline{\psi_\tau^\epsilon}(r, R') \qquad (16.24)$$

which defines a positive, trace-class operator in $L^2(\mathbb{R}^2)$ with $Tr\,\hat{\rho}_\tau^\epsilon = 1$.

We also introduce the approximate reduced density matrix

$$\hat{\rho}_\tau^e(R, R') \equiv \int dr \rho_\tau^e(r, R, r, R') \equiv \int dr \psi_\tau^e(r, R)\overline{\psi_\tau^e}(r, R')$$

$$= \tfrac{1}{2}\left(e^{-i\tau H_0} f_\sigma^+\right)(R)\left(e^{i\tau H_0}\overline{f_\sigma^+}\right)(R') + \tfrac{1}{2}\left(e^{-i\tau H_0} f_\sigma^-\right)(R)\left(e^{i\tau H_0}\overline{f_\sigma^-}\right)(R')$$

$$+ \frac{\Lambda}{2}\left(e^{-i\tau H_0} f_\sigma^+\right)(R)\left(e^{i\tau H_0}\overline{f_\sigma^-}\right)(R') + \frac{\overline{\Lambda}}{2}\left(e^{-i\tau H_0} f_\sigma^-\right)(R)\left(e^{i\tau H_0}\overline{f_\sigma^+}\right)(R')$$

$$\qquad (16.25)$$

where, in the last equality, we took into account that

$$\int dr |g_\mathcal{T}(\epsilon^{-1}\tau, r) + g_\mathcal{R}^{-R_0}(\epsilon^{-1}\tau, r)|^2 = \int dr |g_\mathcal{T}(\epsilon^{-1}\tau, r) + g_\mathcal{R}^{R_0}(\epsilon^{-1}\tau, r)|^2 = 1$$

$$\qquad (16.26)$$

and we defined

$$\Lambda \equiv \int dr \left(g_\mathcal{T}(\epsilon^{-1}\tau, r) + g_\mathcal{R}^{-R_0}(\epsilon^{-1}\tau, r)\right)\left(\overline{g_\mathcal{T}}(\epsilon^{-1}\tau, r) + \overline{g_\mathcal{R}^{R_0}}(\epsilon^{-1}\tau, r)\right).$$

$$\qquad (16.27)$$

Notice that $|\Lambda| \leq 1$ and the case $\Lambda = 1$ can only occur if $\alpha = 0$, i.e., when the interaction is absent.

Since L^2-convergence of the wave function implies convergence of the corresponding density matrix in the trace-class norm, we conclude that $\hat{\rho}_\tau^e$ is a good approximation in the trace-class norm of $\hat{\rho}_\tau^\epsilon$ under the assumptions of Theorem 1 and Proposition 2.

For $\alpha = 0$ we see from (16.25), (16.27) that $\hat{\rho}_\tau^e$ reduces to the pure state describing the coherent superposition of the two wave packets evolving in time according to the free hamiltonian.

Notice that the last two terms in (16.25), usually called nondiagonal terms, are responsible for the interference effects observed when the two wave packets have a nonempty overlapping.

The occurrence of the interference makes the state highly nonclassical and it is the distinctive character of a quantum system with respect to a classical one.

If we switch on the interaction, i.e., for $\alpha > 0$, the effect of the light particle is to reduce the nondiagonal terms by the factor Λ and this means that the interference effects for the heavy particle are correspondingly reduced.

In this sense one can say that there is a (partial) decoherence effect induced on the heavy particle which is completely characterized by the parameter Λ.

In the limit $\Lambda \to 0$, i.e., for $\alpha \to \infty$, the nondiagonal terms and the interference effects are completely cancelled and then the state becomes a classical statistical mixture of the two pure states $e^{-i\tau H_0} f_\sigma^+$, $e^{-i\tau H_0} f_\sigma^-$.

We conclude this note showing that the parameter Λ can be approximated by a simpler expression under further conditions on the parameters.

First we notice that using the first assumption in (16.17) one easily gets

$$\Lambda = \int dr |g_{\mathcal{T}}(\epsilon^{-1}\tau, r)|^2 + \int dr g_{\mathcal{R}}^{-R_0}(\epsilon^{-1}\tau, r)\overline{g_{\mathcal{R}}^{R_0}}(\epsilon^{-1}\tau, r) + O((\delta q_0)^{-n}) .$$

$$(16.28)$$

Moreover we can show that for a large separation of the two wave packets the second integral in (16.28) becomes negligible due to the rapid oscillations of the integrand. More precisely we assume

$$d \gg \frac{1}{\alpha}, \qquad d \gg \delta, \qquad d \equiv 2R_0 . \qquad (16.29)$$

From (16.16), (16.14) we have

$$\mathcal{I} \equiv \int dr g_{\mathcal{R}}^{-R_0}(\epsilon^{-1}\tau, r)\overline{g_{\mathcal{R}}^{R_0}}(\epsilon^{-1}\tau, r)$$

$$= \int dk |\mathcal{R}_\alpha(k)|^2 |\tilde{g}_\delta(-k)|^2 e^{2idk}$$

$$= \int dz \frac{\alpha^2}{\alpha^2 + (\delta^{-1}z - q_0)^2} |\tilde{g}(z)|^2 e^{2id(\delta^{-1}z - q_0)} . \qquad (16.30)$$

Integrating by parts in (16.30) one obtains

$$|\mathcal{I}| = \left| \int dz \frac{\alpha^2}{\alpha^2 + (\delta^{-1}z - q_0)^2} |\tilde{g}(z)|^2 \frac{1}{2id\delta^{-1}} \frac{d}{dz} e^{2id(\delta^{-1}z - q_0)} \right|$$

$$\leq \frac{\delta}{2d} \int dz \left| \frac{d}{dz} \left(\frac{\alpha^2}{\alpha^2 + (\delta^{-1}z - q_0)^2} |\tilde{g}(z)|^2 \right) \right|$$

$$\leq \frac{\delta}{d} \int dz \frac{\alpha^2 \delta^{-1} |\delta^{-1}z - q_0|}{[\alpha^2 + (\delta^{-1}z - q_0)^2]^2} |\tilde{g}(z)|^2$$

$$+ \frac{\delta}{d} \int dz \frac{\alpha^2}{\alpha^2 + (\delta^{-1}z - q_0)^2} |\tilde{g}(z)||\tilde{g}'(z)|$$

$$\leq \frac{1}{d\alpha} + \frac{\delta}{d} \|\tilde{g}'\|, \qquad (16.31)$$

where, in the last line, we have used the trivial estimates

$$\frac{\alpha^2 \delta^{-1}|\xi|}{(\alpha^2 + \xi^2)^2} \leq \frac{\alpha^2}{\alpha^2 + \xi^2}\frac{\delta^{-1}|\xi|}{\alpha^2 + \xi^2} \leq \frac{\delta^{-1}|\xi|}{\alpha^2 + \xi^2} \leq \frac{1}{\delta\alpha} \qquad (16.32)$$

and the Schwartz inequality.

Using the estimate (16.31) in (16.28) we conclude that

$$\left| \Lambda - \int dr |g_{\mathcal{T}}(\epsilon^{-1}\tau, r)|^2 \right| \leq \frac{1}{d\alpha} + \frac{\delta}{d}\,\|\tilde{g}'\| + O((\delta q_0)^{-n})\,. \qquad (16.33)$$

This means that, under the assumptions (16.17),(16.29), the decoherence effect is measured by the transmission probability of the light particle which is explicitly given by

$$\int dr |g_{\mathcal{T}}(\epsilon^{-1}\tau, r)|^2 = \int dk |\mathcal{T}_\alpha(k)|^2 |\tilde{g}_\delta(k)|^2 \qquad (16.34)$$

The previous analysis of decoherence induced by scattering is clearly limited by the consideration of a two-body system.

For a more satisfactory treatment one should consider a model with N light particles scattered by the heavy one. The expected result is that the effect of the scattering events is cumulative and then the decoherence effect is increased at each step ([JZ]).

We observe that a rigorous proof of this fact would require good estimates for the wave operator in a case of many-body scattering problem and it is a nontrivial open question.

References

[AGH-KH] Albeverio S., Gesztesy F., Hoegh-Krohn R., Holden H., *Solvable Models in Quantum Mechanics*, Springer-Verlag, 1988.

[BGJKS] Blanchard Ph., Giulini D., Joos E., Kiefer C., Stamatescu I.-O., *Decoherence: Theoretical, Experimental and Conceptual Problems*. Lect. Notes in Phys. Vol. 538, Springer, 2000.

[D] Dell'Antonio G., *On Decoherence*, Preprint, 2003.

[DFT] Duerr D., Figari R., Teta A., Decoherence in a two-particle model. http://www.ma.utexas.edu 02-442, submitted to *J. Math. Phys.*

[GF] Gallis M.R., Fleming G.N., Environmental and spontaneous localization, *Phys. Rev.* **A42** (1990), 38-48.

[GJKKSZ] Giulini D., Joos E., Kiefer C., Kupsch J., Stamatescu I.-O., Zeh H.D., *Decoherence and the Appearance of a Classical World in Quantum Theory*, Springer, 1996.

[JZ] Joos E., Zeh H.D., The emergence of classical properties through interaction with the environment. *Z. Phys.* **B59** (1985), 223–243.

[S] Schulman L.S., Application of the propagator for the delta function potential. In: *Path Integrals from mev to Mev*, Gutzwiller M.C., Ioumata A., Klauder J.K., Streit L., eds., World Scientific, 1986.

[T] Tegmark M., Apparent wave function collapse caused by scattering. *Found. Phys. Lett.*, **6** (1993), 571–590.

Propagation of Wigner Functions for the Schrödinger Equation with a Perturbed Periodic Potential

S. Teufel and G. Panati

Zentrum Mathematik
Technische Universität München
D-80333 München

Summary. Let V_Γ be a lattice periodic potential and A and ϕ external electromagnetic potentials which vary slowly on the scale set by the lattice spacing. It is shown that the Wigner function of a solution of the Schrödinger equation with Hamiltonian operator $H = \frac{1}{2}(-\mathrm{i}\nabla_x - A(\varepsilon x))^2 + V_\Gamma(x) + \phi(\varepsilon x)$ propagates along the flow of the semiclassical model of solid states physics up to an error of order ε. If ε-dependent corrections to the flow are taken into account, the error is improved to order ε^2. We also discuss the propagation of the Wigner measure. The results are obtained as corollaries of an Egorov type theorem proved in [PST3].

17.1 Introduction

One of the central questions of solid state physics is to understand the motion of electrons in the periodic potential which is generated by the ionic cores. While this problem is quantum mechanical, many electronic properties of solids can be understood already in the semiclassical approximation [AsMe, Ko, Za]. One argues that for suitable wave packets, which are spread over many lattice spacings, the main effect of a periodic potential V_Γ on the electron dynamics corresponds to changing the dispersion relation from the free kinetic energy $E_{\text{free}}(p) = \frac{1}{2}p^2$ to the modified kinetic energy $E_n(p)$ given by the nth Bloch function. Otherwise the electron responds to slowly varying external potentials A, ϕ as in the case of a vanishing periodic potential. Thus the semiclassical equations of motion are

$$\dot{r} = \nabla E_n(\kappa)\,, \qquad \dot{\kappa} = -\nabla\phi(r) + \dot{r} \times B(r)\,, \tag{17.1}$$

where $\kappa = k - A(r)$ is the kinetic momentum and $B = \operatorname{curl} A$ is the magnetic field. (We choose units in which the Planck constant $\hbar$, the speed c of light, and the mass m of the electron are equal to 1, and absorb the charge e into the potentials.) The corresponding equations of motion for the canonical variables (r, k) are generated by the Hamiltonian

$$H_{\mathrm{sc}}(r, k) = E_n\big(k - A(r)\big) + \phi(r) \,, \tag{17.2}$$

where r is the position and k the quasi-momentum of the electron. Note that there is a semiclassical evolution for each Bloch band separately. The distinction between the canonical variable k, the Bloch- or quasi-momentum, and the kinetic momentum $\kappa = k - A(r)$ is often not made explicit in the physics literature. It is, however, crucial for the formulation of the precise connection between the semiclassical equations of motion (17.1) and the underlying Schrödinger equation (17.4).

In [PST$_3$] we use adiabatic perturbation theory in order to understand on a mathematical level how these semiclassical equations emerge from the underlying Schrödinger equation

$$\mathrm{i}\,\partial_s\,\psi(y, s) = \left(\tfrac{1}{2}\big(-\mathrm{i}\nabla_y - A(\varepsilon y)\big)^2 + V_\Gamma(y) + \phi(\varepsilon y) \right) \psi(y, s) \tag{17.3}$$

in the limit $\varepsilon \to 0$ at leading order. In addition, the order ε corrections to (17.1) are established, see Equation (17.7).

In (17.3) the potential $V_\Gamma : \mathbb{R}^d \to \mathbb{R}$ is periodic with respect to some regular lattice Γ generated through the basis $\{\gamma_1, \ldots, \gamma_d\}$, $\gamma_j \in \mathbb{R}^d$, i.e.,

$$\Gamma = \left\{ x \in \mathbb{R}^d : x = \textstyle\sum_{j=1}^{d}\alpha_j\,\gamma_j \ \text{ for some } \alpha \in \mathbb{Z}^d \right\}$$

and $V_\Gamma(\cdot + \gamma) = V_\Gamma(\cdot)$ for all $\gamma \in \Gamma$. The lattice spacing defines the microscopic spatial scale. The external potentials $A(\varepsilon y)$ and $\phi(\varepsilon y)$, with $A : \mathbb{R}^d \to \mathbb{R}^d$ and $\phi : \mathbb{R}^d \to \mathbb{R}$, are slowly varying on the scale of the lattice, as expressed through the dimensionless scale parameter ε, $\varepsilon \ll 1$. In particular, this means that the external fields are weak compared to the fields generated by the ionic cores, a condition which is satisfied for real metals even for the strongest external electrostatic fields available and for a wide range of magnetic fields, cf. [AsMe], Chapter 12.

Note that the external forces due to A and ϕ are of order ε and therefore have to act over a time of order ε^{-1} to produce finite changes, which is taken as the definition of the macroscopic time scale. Hence, one is interested in solutions of (17.3) for macroscopic times. The macroscopic space-time scale (x, t) is defined through $x = \varepsilon y$ and $t = \varepsilon s$. With this change of variables Equation (17.3) reads

$$\mathrm{i}\varepsilon\,\partial_t\,\psi^\varepsilon(x, t) = \left(\tfrac{1}{2}\big(-\mathrm{i}\varepsilon\nabla_x - A(x)\big)^2 + V_\Gamma(x/\varepsilon) + \phi(x) \right) \psi^\varepsilon(x, t) \tag{17.4}$$

with initial conditions $\psi^\varepsilon(x) = \varepsilon^{-d/2}\psi(x/\varepsilon)$. If $V_\Gamma = 0$, then the limit $\varepsilon \to 0$ in Equation (17.4) is the usual semiclassical limit with ε replacing $\hbar$.

The problem of deriving (17.1) from the Schrödinger equation (17.3) in the limit $\varepsilon \to 0$ has been attacked along several routes. In the physics literature (17.1) is usually accounted for by constructing suitable semiclassical wave packets. We refer to [Lu, Ko, Za]. The few mathematical approaches to the time-dependent problem (17.4) extend techniques from semiclassical analysis, as the Gaussian beam construction [GRT, DGR], or Wigner measures [GMMP, BFPR, BMP].

In this article we explain and elaborate on recent results from [PST$_3$]. In [PST$_3$] we derived (17.1) from (17.4) for quite general external potentials A and ϕ. The construction is based on the space-adiabatic perturbation theory developed in [PST$_1$, PST$_2$, Te], see also [NeSo] and the contribution of G. Nenciu in the present volume. The crucial observation is that the step from (17.3) to (17.1) involves actually two approximations. Semiclassical behavior can only emerge if a Bloch band is separated by a gap from the other bands and thus the corresponding subspace decouples adiabatically from its orthogonal complement. The dynamics inside this adiabatic subspace is governed by an effective Hamiltonian $\widehat{h}^\varepsilon_{\mathrm{eff}}$, which is explicitly given as an ε-pseudo-differential operator. Eventually, the semiclassical limit of $\widehat{h}^\varepsilon_{\mathrm{eff}}$ leads to (17.1).

Hence (17.3) needs to be reformulated as a space-adiabatic problem. This has been done first in [HST] for the case of zero magnetic field and then in [PST$_3$] for general electric and magnetic fields. The results obtained in this way constitute not only the derivation of the semiclassical model (17.1) in this generality, but they allow us to compute systematically higher order corrections in the small parameter ε. It turns out that the electron acquires a k-dependent electric moment $\mathcal{A}_n(k)$ and magnetic moment $\mathcal{M}_n(k)$. If the nth band is nondegenerate (and isolated) with Bloch eigenfunctions $\psi_n(k,x)$, the electric dipole moment is given by the Berry connection

$$\mathcal{A}_n(k) = \mathrm{i}\left\langle \psi_n(k), \nabla\psi_n(k)\right\rangle \tag{17.5}$$

and the magnetic moment by the Rammal–Wilkinson phase

$$\mathcal{M}_n(k) = \tfrac{\mathrm{i}}{2}\left\langle \nabla\psi_n(k),\ \times(H_{\mathrm{per}}(k) - E(k))\nabla\psi_n(k)\right\rangle. \tag{17.6}$$

Here $\langle\,\cdot\,,\,\cdot\,\rangle$ is the inner product in $L^2(\mathbb{R}^d/\Gamma)$ and $H_{\mathrm{per}}(k)$ is H of (17.3) with $\phi = 0 = A$ for fixed Bloch momentum k. Note that E_n, $\mathcal{A}_n$ and $\mathcal{M}_n$ are Γ^*-periodic functions of k, where Γ^* is the lattice dual to Γ. Hence one can as well think of them as functions on the domain $M^* = \mathbb{R}^d/\Gamma^*$, the first Brillouin zone.

The semiclassical equations of motion including first order corrections read

$$\dot{r} = \nabla_\kappa\Big(E_n(\kappa) - \varepsilon\, B(r)\cdot\mathcal{M}_n(\kappa)\Big) - \varepsilon\,\dot{k}\times\Omega_n(\kappa)\,,$$

$$\dot{\kappa} = -\nabla_r\Big(\phi(r) - \varepsilon\,B(r)\cdot\mathcal{M}_n(\kappa)\Big) + \dot{r}\times B(r)$$

with $\Omega_n(k) = \nabla\times\mathcal{A}_n(k)$ the curvature of the Berry connection.

In order to state the precise connection between the semiclassical equations of motion (17.1) resp. their refined version (17.7) and the underlying Schrödinger equation (17.4), we need some more notation. Let

$$H^\varepsilon = \tfrac{1}{2}\big(-\mathrm{i}\varepsilon\nabla_x - A(x)\big)^2 + V_\Gamma(x/\varepsilon) + \phi(x) \tag{17.7}$$

be the Hamiltonian of (17.4). Under the following assumption on the potentials, which will be imposed throughout, H^ε is self-adjoint on $H^2(\mathbb{R}^d)$. Here $C_{\mathrm{b}}^\infty(\mathbb{R}^d)$ denotes the space of smooth functions which are bounded together with all their derivatives.

Assumption 1 *Let V_Γ be infinitesimally bounded with respect to $-\Delta$ and assume that $\phi \in C_{\mathrm{b}}^\infty(\mathbb{R}^d, \mathbb{R})$ and $A_j \in C_{\mathrm{b}}^\infty(\mathbb{R}^d, \mathbb{R})$ for any $j \in \{1, \ldots, d\}$.*

To each isolated Bloch band E_n there corresponds an associated almost invariant band-subspace $\Pi_n^\varepsilon L^2(\mathbb{R}^d)$. The orthogonal projector Π_n^ε onto this subspace is constructed in [PST$_3$]. Only for states which start in this subspace and thus, by construction, remain there up to small errors, can the semiclassical equations of motion (17.7) have any significance.

The flow of the dynamical system (17.7) is denoted by $\Phi_\varepsilon^t : \mathbb{R}^{2d} \to \mathbb{R}^{2d}$ or in canonical coordinates $(r, k) = (r, \kappa + A(r))$ by

$$\overline{\Phi}_\varepsilon^t(r, k) = \Big(\Phi_{\varepsilon\,r}^t\big(r, k - A(r)\big),\ \Phi_{\varepsilon\,\kappa}^t\big(r, k - A(r)\big) + A(r)\Big).$$

The existence of the smooth family of diffeomorphisms Φ_ε^t is not completely obvious from (17.7) alone, but follows from the Hamiltonian formulation of (17.7) presented in the next section.

Notation 1 *Throughout this paper we will use the Fréchet space*

$$\mathcal{C} = C_{\mathrm{b}}^\infty(\mathbb{R}^{2d}),$$

equipped with the metric $d_\mathcal{C}$ induced by the standard family of semi-norms

$$\|a\|_\alpha = \|\partial^\alpha a\|_\infty, \quad \alpha \in \mathbb{N}_0^{2d},$$

and the subspace of Γ^*-periodic observables

$$\mathcal{C}_{\mathrm{per}} = \{a \in \mathcal{C} : a(r, k + \gamma^*) = a(r, k)\ \forall \gamma^* \in \Gamma^*\}.$$

We abbreviate $d_\mathcal{C}(a) := d_\mathcal{C}(a, 0)$. $\square$

The main result of [PST$_3$] on the semiclassical limit of (17.4) is the following Egorov type theorem.

Theorem 1. *Let E_n be an isolated, nondegenerate Bloch band. For each finite time-interval $I \subset \mathbb{R}$ there is a constant $C < \infty$, such that for all $a \in \mathcal{C}_{\mathrm{per}}$ with Weyl quantization $\widehat{a} = a(x, -\mathrm{i}\varepsilon\nabla_x)$ one has*

$$\left\| \Big(\mathrm{e}^{\mathrm{i}H^\varepsilon t/\varepsilon}\, \widehat{a}\, \mathrm{e}^{-\mathrm{i}H^\varepsilon t/\varepsilon} - \widehat{a \circ \overline{\Phi}_0^t} \Big)\, \Pi_n^\varepsilon \right\|_{\mathcal{B}(L^2(\mathbb{R}^d))} \leq \varepsilon\, C\, d_\mathcal{C}(a) \tag{17.8}$$

and

$$\left\| \Pi_n^\varepsilon \Big(\mathrm{e}^{\mathrm{i}H^\varepsilon t/\varepsilon}\, \widehat{a}\, \mathrm{e}^{-\mathrm{i}H^\varepsilon t/\varepsilon} - \widehat{a \circ \overline{\Phi}_\varepsilon^t} \Big)\, \Pi_n^\varepsilon \right\|_{\mathcal{B}(L^2(\mathbb{R}^d))} \leq \varepsilon^2\, C\, d_\mathcal{C}(a). \tag{17.9}$$

Remark 1. The corresponding statement in [PST$_3$] does not make explicit the dependence of the error on the observable a. However, the more precise version formulated here is a standard consequence of the Calderon–Vaillancourt theorem and the fact that composition with $\overline{\Phi}^t_\varepsilon$ is a continuous map from $\mathcal{C}$ into itself.   □

Remark 2. On an abstract level the distinction between the functions Φ^t_ε and $\overline{\Phi}^t_\varepsilon$ is immaterial, since both functions express the same dynamical flow in two systems of coordinates. However, the distinction between the systems of coordinates becomes important when the quantization is considered. The Weyl quantization appearing in (17.8) and (17.9) must be understood with respect to the system of coordinates (r, k). Analogous considerations hold true for formulas involving a Wigner transform, as in Corollary 1.   □

The main objective of this article is to elaborate on Theorem 1 in order to make contact to alternative approaches and results on the semiclassical limit of (17.4). These are, as mentioned above, Wigner functions [GMMP, BFPR, BMP], semiclassical wave packets [Lu, Ko, Za, SuNi] and WKB-type solutions of (17.4) [Bu, GRT, DGR]. We focus on the semiclassical transport of Wigner functions and Wigner measures in the following. Before we do so, it is worthwhile to first examine the equations of motion (17.7) in some more detail.

17.2 The refined semiclassical equations of motion

The dynamical equations (17.7), which define the ε-corrected semiclassical model, can be written as

$$\dot{r} = \nabla_\kappa H_{\mathrm{sc}}(r, \kappa) - \varepsilon\, \dot{\kappa} \times \Omega_n(\kappa)\,,$$

$$\dot{\kappa} = -\nabla_r H_{\mathrm{sc}}(r, \kappa) + \dot{r} \times B(r) \tag{17.10}$$

with

$$H_{\mathrm{sc}}(r, \kappa) := E_n(\kappa) + \phi(r) - \varepsilon\, \mathcal{M}_n(k) \cdot B(r)\,. \tag{17.11}$$

We shall show that (17.10) are the Hamiltonian equations of motion for (17.11) with respect to a suitable ε-dependent symplectic form $\Theta_{B,\varepsilon}$. The semiclassical equations of motion (17.7) are defined for arbitrary dimension d. However, to simplify presentation, we use a notation motivated by the vector product and the duality between 1-forms and 2-forms for $d = 3$, which we briefly explain.

Notation 2 *If $d \neq 3$, then B, Ω_n and $\mathcal{M}_n$ are 2-forms with components*

$$B_{ij}(r) = \partial_i A_j(r) - \partial_j A_i(r)\,, \qquad \Omega_{ij}(k) = \partial_i \mathcal{A}_j(k) - \partial_j \mathcal{A}_i(k)$$

and

$$\mathcal{M}_{ij}(k) = \mathrm{Re}\,\tfrac{\mathrm{i}}{2}\,\langle \partial_i \psi_n(k), (H_{\mathrm{per}} - E)(k)\,\partial_j \psi_n(k)\rangle\,.$$

For $d = 3$. a 2-form $B_{ij}(r)$ is naturally associated with the vector $B_k(r) = \epsilon_{kij}\,B_{ij}(r)$. We use the convention that summation over repeated indices is implicit. Then in (17.7) the inner product $B \cdot \mathcal{M}_n$ refers to the product of the associated vectors and we generalize the notation to arbitrary dimension d using the inner product of 2-forms defined through

$$B \cdot \mathcal{M} := *^{-1}(B \wedge *\mathcal{M}) = \sum_{j=1}^{d}\sum_{i=1}^{d} B_{ij}\mathcal{M}_{ij}\,,$$

where $*$ denotes the Hodge duality induced by the euclidean metric. In the same spirit for a vector field w and a 2-form F, the generalized "vector product" is

$$(w \times F)_j := (*^{-1}(w \wedge *F))_j = \sum_{i=1}^{d} w_i F_{ij}\,,$$

where the duality between 1-forms and vector fields is used implicitly. $\square$

We keep fixed the system of coordinates $z = (r, \kappa)$ in $\mathbb{R}^{2d}$ for the following. The standard symplectic form $\Theta_0 = \Theta_0(z)_{lm}\,\mathrm{d}z_m \wedge \mathrm{d}z_l$, where $l, m \in \{1, \ldots, 2d\}$, has coefficients given by the constant matrix

$$\Theta_0(z) = \begin{pmatrix} 0 & -\mathbb{I} \\ \mathbb{I} & 0 \end{pmatrix}\,,$$

where $\mathbb{I}$ is the identity matrix in $\mathrm{Mat}(d, \mathbb{R})$. The symplectic form, which turns (17.10) into Hamilton's equation of motion for H_{sc}, is given by the 2-form $\Theta_{B,\varepsilon} = \Theta_{B,\varepsilon}(z)_{lm}\,\mathrm{d}z_m \wedge \mathrm{d}z_l$ with coefficients

$$\Theta_{B,\varepsilon}(r, \kappa) = \begin{pmatrix} B(r) & -\mathbb{I} \\ \mathbb{I} & \varepsilon\,\Omega_n(\kappa) \end{pmatrix}\,. \tag{17.12}$$

For $\varepsilon = 0$ the 2-form $\Theta_{B,\varepsilon}$ coincides with the magnetic symplectic form Θ_B usually employed to describe in a gauge-invariant way the motion of a particle in a magnetic field ([MaRa], Section 6.6). For ε small enough, the matrix (17.12) defines a symplectic form, i.e., a closed nondegenerate 2-form.

With these definitions the corresponding Hamiltonian equations are

$$\Theta_{B,\varepsilon}(z)\,\dot{z} = \mathrm{d}H_{\mathrm{sc}}(z)\,,$$

or equivalently

$$\begin{pmatrix} B(r) & -\mathbb{I} \\ \mathbb{I} & \varepsilon\,\Omega_n(\kappa) \end{pmatrix} \begin{pmatrix} \dot{r} \\ \dot{\kappa} \end{pmatrix} = \begin{pmatrix} \nabla_r H(r, \kappa) \\ \nabla_\kappa H(r, \kappa) \end{pmatrix}\,,$$

which agrees with (17.10). We notice that this discussion remains valid if Ω_n admits a potential only locally, as it happens generically for magnetic Bloch bands.

The symplectic structure is therefore determined by the magnetic field $B(r)$ and by the curvature of the Berry connection $\Omega(k)$, which encodes relevant information about the geometry of the Bloch bundle $\psi_n(k, \cdot) \mapsto k \in M^*$. One can show that, whenever the Hamiltonian H_{per} has time-reversal symmetry one has that $\Omega_n(-k) = -\Omega_n(k)$. Moreover, if the lattice Γ has a center of inversion, then $\Omega_n(-k) = \Omega_n(k)$. Thus, the two symmetries together imply that $\Omega_n(k)$ vanishes pointwise. But there are many crystals which do *not* have a center of inversion and, more important, in the presence of a strong uniform magnetic field the time-reversal symmetry is broken. The latter is the typical setup to describe the Quantum Hall Effect, a situation in which the curvature of the Berry connection plays a prominent role. Indeed, the equations of motion (17.7) provide a simple semiclassical explanation of the Quantum Hall Effect. Let us specialize (17.7) to two dimensions and take $B(r) = 0$, $\phi(r) = -\mathcal{E} \cdot r$, i.e., a weak driving electric field and a strong uniform magnetic field with rational flux. Then, since $\kappa = k$, the equations of motion become $\dot{r} = \nabla_k E_n(k) + \mathcal{E}^\perp \Omega_n(k)$, $\dot{k} = \mathcal{E}$, where Ω_n is now scalar, and $\mathcal{E}^\perp$ is $\mathcal{E}$ rotated by $\pi/2$. We assume initially $k(0) = k$ and a completely filled band, which means to integrate with respect to k over the first Brillouin zone M^*. Then the average current for band n is given by

$$j_n = \int_{M^*} \mathrm{d}k\, \dot{r}(k) = \int_{M^*} \mathrm{d}k\, \left(\nabla_k E_n(k) - \mathcal{E}^\perp \Omega_n(k)\right) = -\mathcal{E}^\perp \int_{M^*} \mathrm{d}k\, \Omega_n(k).$$

$\int_{M^*} \mathrm{d}k\, \Omega_n(k)$ is the Chern number of the magnetic Bloch bundle and as such an integer, cf. [TKNN]. Further applications related to the semiclassical first order corrections are the anomalous Hall effect [JNM] and the thermodynamics of the Hofstadter model [GaAv].

17.3 Semiclassical transport of Wigner functions

Theorem 1 provides a semiclassical description of the evolution of observables. The most direct way to turn it into a description for the semiclassical evolution of states is via duality, i.e., via the Wigner function. Recall that according to the Calderon–Vaillancourt theorem there is a constant $C < \infty$ depending only on the dimension d such that for $a \in \mathcal{C}$ one has

$$|\langle \psi, \widehat{a}\, \psi\rangle_{L^2(\mathbb{R}^d)}| \leq C\, d_{\mathcal{C}}(a)\, \|\psi\|^2. \tag{17.13}$$

Hence, the map $\mathcal{C} \ni a \mapsto \langle \psi, \widehat{a}\, \psi\rangle \in \mathbb{C}$ is continuous and thus defines an element w_ε^ψ of the dual space $\mathcal{C}'$, the Wigner function of ψ. Writing

$$\langle \psi, \widehat{a}\, \psi\rangle =: \langle w_\varepsilon^\psi, a\rangle_{\mathcal{C}',\mathcal{C}} =: \int_{\mathbb{R}^{2d}} \mathrm{d}q\, \mathrm{d}p\, a(q, p)\, w_\varepsilon^\psi(q, p) \tag{17.14}$$

and inserting into (17.14) the definition of the Weyl quantization for $a \in \mathcal{S}(\mathbb{R}^{2d})$,

$$(\widehat{a}\psi)(x) = \frac{1}{(2\pi)^d} \int_{\mathbb{R}^{2d}} \mathrm{d}\xi\, \mathrm{d}y\, a\big(\tfrac{1}{2}(x+y), \varepsilon\xi\big)\, \mathrm{e}^{\mathrm{i}\xi\cdot(x-y)}\, \psi(y)\,,$$

one arrives at the formula

$$w_\varepsilon^\psi(q,p) = \frac{1}{(2\pi)^d} \int_{\mathbb{R}^d} \mathrm{d}\xi\, \mathrm{e}^{\mathrm{i}\xi\cdot p}\, \psi^*(q + \varepsilon\xi/2)\, \psi(q - \varepsilon\xi/2) \qquad (17.15)$$

for the Wigner function. Direct computation yields

$$\| w_\varepsilon^\psi \|_{L^2(\mathbb{R}^{2d})} = \varepsilon^{-d}\, (2\pi)^{-d/2}\, \| \psi \|_{L^2(\mathbb{R}^d)}^2\,.$$

Therefore, $w_\varepsilon^\psi \in L^2(\mathbb{R}^{2d})$ for all $\varepsilon > 0$, which explains the notion of Wigner function. Although w_ε^ψ is obviously real-valued, it attains also negative values in general. Hence, it does not define a probability distribution on phase space. However, it correctly produces quantum mechanical distributions via (17.14).

With this preparation we obtain the following corollary of Theorem 1, which says that the Wigner function of the solution of the Schrödinger equation (17.4) is approximately transported along the classical flow of (17.1) resp. (17.7).

Corollary 1. *Let E_n be an isolated, nondegenerate Bloch band. Then for each finite time-interval $I \subset \mathbb{R}$ there is a constant $C < \infty$ such that for $t \in I$, $u \in \mathcal{C}_{\mathrm{per}}$ and for $\psi_0 \in \Pi_n^\varepsilon L^2(\mathbb{R}^d)$ one has*

$$\left| \left\langle \left(w_\varepsilon^{\psi_t} - w_\varepsilon^{\psi_0} \circ \overline{\Phi}_0^{-t} \right), a \right\rangle_{\mathcal{C}',\mathcal{C}} \right| \leq \varepsilon\, C\, d_{\mathcal{C}}(a)\, \|\psi_0\|^2$$

and

$$\left| \left\langle \left(w_\varepsilon^{\psi_t} - w_\varepsilon^{\psi_0} \circ \overline{\Phi}_\varepsilon^{-t} \right), a \right\rangle_{\mathcal{C}',\mathcal{C}} \right| \leq \varepsilon^2\, C\, d_{\mathcal{C}}(a)\, \|\psi_0\|^2\,.$$

Here $\psi_t = \mathrm{e}^{-\mathrm{i}H^\varepsilon t/\varepsilon}\psi_0$ is the solution of the Schrödinger equation (17.4).

Remark 3. When proving results for the transport of Wigner functions or Wigner measures it is common, e.g., [GMMP, MMP, BMP], to write down the transport equation for $w_\varepsilon(t) := w_\varepsilon^{\psi_0} \circ \overline{\Phi}_0^{-t}$ instead of using the flow $\overline{\Phi}_0^t$. Clearly our results can be reformulated in this way, cf. Corollary 2, but the resulting transport equation looks complicated compared to the simple dynamical system (17.1) governing its characteristics. $\qquad\qquad\square$

Proof (Proof of Corollary 1). The result is rather a reformulation of Theorem 1 than a real corollary. According to the Definition (17.14) and Theorem 1 one has

$$\langle w_\varepsilon^{\psi_t}, a\rangle_{\mathcal{C}',\mathcal{C}} = \langle \psi_t, \widehat{a}\,\psi_t\rangle_{L^2(\mathbb{R}^d)}$$
$$= \langle \psi_0, \mathrm{e}^{\mathrm{i}H^\varepsilon t/\varepsilon}\,\widehat{a}\,\mathrm{e}^{-\mathrm{i}H^\varepsilon t/\varepsilon}\,\psi_0\rangle_{L^2(\mathbb{R}^d)}$$
$$= \langle \psi_0, \varPi_n^\varepsilon\,\mathrm{e}^{\mathrm{i}H^\varepsilon t/\varepsilon}\,\widehat{a}\,\mathrm{e}^{-\mathrm{i}H^\varepsilon t/\varepsilon}\,\varPi_n^\varepsilon\,\psi_0\rangle_{L^2(\mathbb{R}^d)}$$
$$= \langle \psi_0, \varPi_n^\varepsilon\,\widehat{a\circ\overline{\varPhi}_\varepsilon^{\,t}}\,\varPi_n^\varepsilon\,\psi_0\rangle_{L^2(\mathbb{R}^d)} + \mathcal{O}(\varepsilon^2)$$
$$= \langle \psi_0, \widehat{a\circ\overline{\varPhi}_\varepsilon^{\,t}}\,\psi_0\rangle_{L^2(\mathbb{R}^d)} + \mathcal{O}(\varepsilon^2)\,.$$

Since the map $\mathcal{C}\ni a \mapsto a\circ\overline{\varPhi}_\varepsilon^{\,t} \in \mathcal{C}$ is continuous, the duality relation (17.14) can be applied again and yields

$$\langle\psi_0, \widehat{a\circ\overline{\varPhi}_\varepsilon^{\,t}}\,\psi_0\rangle_{L^2(\mathbb{R}^d)} = \left\langle w_\varepsilon^{\psi_0}, a\circ\overline{\varPhi}_\varepsilon^{\,t}\right\rangle_{\mathcal{C}',\mathcal{C}} = \left\langle w_\varepsilon^{\psi_0}\circ\overline{\varPhi}_\varepsilon^{\,-t}, a\right\rangle_{\mathcal{C}',\mathcal{C}}\,.$$

Since the functions E_n, $\mathcal{M}_n$ and Ω_n appearing in the equations of motion (17.7) are all Γ^* periodic, the natural phase space for the flow (17.7) is $\mathbb{R}^d\times\mathbb{T}^*$ rather than $\mathbb{R}^{2d}$. Here $\mathbb{T}^d := \mathbb{R}^d/\Gamma^*$ is the first Brillouin zone M^* equipped with periodic boundary conditions. Hence one can fold the Wigner transform onto the first Brillouin zone and define

$$w_{\varepsilon\,\mathrm{red}}^\psi(r,k) = \sum_{\gamma^*\in\Gamma^*} w_\varepsilon^\psi(r, k+\gamma^*) \quad \text{for} \quad (r,k)\in\mathbb{R}^d\times\mathbb{T}^d\,. \tag{17.16}$$

Then for periodic observables a it follows that

$$\int_{\mathbb{R}^{2d}} \mathrm{d}r\,\mathrm{d}k\, a(r,k)\, w_\varepsilon^\psi(r,k) = \sum_{\gamma^*\in\Gamma^*} \int_{\mathbb{R}^d\times M^*} \mathrm{d}r\,\mathrm{d}k\, a(r, k+\gamma^*)\, w_\varepsilon^\psi(r, k+\gamma^*)$$
$$= \sum_{\gamma^*\in\Gamma^*} \int_{\mathbb{R}^d\times M^*} \mathrm{d}r\,\mathrm{d}k\, a(r, k)\, w_\varepsilon^\psi(r, k+\gamma^*)$$
$$= \int_{\mathbb{R}^d\times\mathbb{T}^d} \mathrm{d}r\,\mathrm{d}k\, a(r, k)\, w_{\varepsilon\,\mathrm{red}}^\psi(r, k)\,.$$

Thus the statement of Corollary 1 in terms of the reduced Wigner function becomes

$$\langle\psi_t, \widehat{a}\,\psi_t\rangle_{L^2(\mathbb{R}^d)} = \int_{\mathbb{R}^d\times\mathbb{T}^d} \mathrm{d}r\,\mathrm{d}k\, a(r,k) \left(w_{\varepsilon\,\mathrm{red}}^{\psi_0}\circ\overline{\varPhi}_\varepsilon^{\,-t}\right)(r,k) + \mathcal{O}(\varepsilon^2)\,.$$

Note that the reduced Wigner function $w_{\varepsilon\,\mathrm{red}}^\psi$ coincides with the "band-Wigner function" of [MMP] and the "Wigner series" of [BMP], both defined as

$$w_{\varepsilon\,\mathrm{s}}^\psi(r,k) = \frac{1}{|M^*|} \sum_{\gamma\in\Gamma} \mathrm{e}^{\mathrm{i}\gamma\cdot k}\, \psi(r + \varepsilon\gamma/2)\, \psi^*(r - \varepsilon\gamma/2)\,.$$

This follows by a simple computation on the dense set $\psi\in\mathcal{S}(\mathbb{R}^d)$:

$$w^{\psi}_{\varepsilon\,\mathrm{red}}(r,k) = \sum_{\gamma^*\in\Gamma^*} w^{\psi}_{\varepsilon}(r, k+\gamma^*)$$

$$= \frac{1}{(2\pi)^d}\sum_{\gamma^*\in\Gamma^*}\int_{\mathbb{R}^d} d\xi\, e^{i\xi\cdot\gamma^*}\, e^{i\xi\cdot k}\,\psi(r+\varepsilon\xi/2)\,\psi^*(r-\varepsilon\xi/2)$$

$$= \frac{1}{|M^*|}\int_{\mathbb{R}^d} d\xi\, \delta_\Gamma(\xi)\, e^{i\xi\cdot k}\,\psi(r+\varepsilon\xi/2)\,\psi^*(r-\varepsilon\xi/2)$$

$$= \frac{1}{|M^*|}\sum_{\gamma\in\Gamma} e^{i\gamma\cdot k}\,\psi(r+\varepsilon\gamma/2)\,\psi^*(r-\varepsilon\gamma/2)\,,$$

where $\delta_\Gamma(\xi) = \sum_{\gamma\in\Gamma}\delta(\xi-\gamma)$. We used the Poisson formula

$$\frac{1}{(2\pi)^d}\sum_{\gamma^*\in\Gamma^*} e^{i\xi\cdot\gamma^*} = \frac{1}{|M^*|}\delta_\Gamma(\xi)\,.$$

17.4 Classical transport of the Wigner measure

We now turn to the Wigner measure. Recall that the Wigner function $w^{\psi}_{\varepsilon}(q,p)$ can be negative and, as a consequence, does not define a probability distribution on phase space. In the limit $\varepsilon \to 0$ however, w^{ψ}_{ε} weakly converges to a positive finite Radon measure $\mu^{\psi} \in \mathcal{M}^+_{\mathrm{b}}(\mathbb{R}^{2d})$ on phase space $\mathbb{R}^{2d}$, the Wigner measure of ψ. For surveys on Wigner measures see e.g., [LiPa, GMMP].

Proposition 1. *Let $\varepsilon_j \overset{j\to\infty}{\to} 0$ and $\{\psi_j\}_{j\in\mathbb{N}} \subset L^2(\mathbb{R}^d)$ be bounded, then $\{w^{\psi_j}_{\varepsilon_j}\}_{j\in\mathbb{N}} \subset \mathcal{C}'$ is weak-$*$ compact and every limit point $\mu \in \mathcal{C}'$ defines a bounded positive Radon measure, called a Wigner measure of $\{\psi_j\}_{j\in\mathbb{N}}$.*

Proof. The Calderon–Vaillancourt theorem (17.13) implies that $\{w^{\psi_j}_{\varepsilon_j}\} \subset \mathcal{C}'$ is bounded. Hence, it is weak-$*$ compact. By (17.14) and the semiclassical sharp Gårding inequality, e.g., Theorem 7.12 in [DiSj], it follows that for each $a \geq 0$ there is some $C < \infty$ such that

$$\langle w^{\psi}_{\varepsilon}, a\rangle_{\mathcal{C}',\mathcal{C}} \geq -C\,\varepsilon\,\|\psi\|^2 \qquad \text{for all } \psi \in L^2(\mathbb{R}^d)\,.$$

This implies the positivity of all limit points in $\mathcal{C}'$, which therefore define measures.

Let $\mu \in \mathcal{C}'$ be such a limit point with, after possible extraction of a subsequence, $w^{\psi_j}_{\varepsilon_j} \overset{*}{\to} \mu$. From (17.14) it follows that

$$\langle w^{\psi}_{\varepsilon}, 1\rangle_{\mathcal{C}',\mathcal{C}} = \|\psi\|^2_{L^2(\mathbb{R}^d)} \qquad \text{for all } \psi \in L^2(\mathbb{R}^d)\,,$$

and thus,

$$\mu(\mathbb{R}^{2d}) = \sup\{\mu(K) : K \subset \mathbb{R}^{2d} \text{ compact}\}$$
$$\leq \langle \mu, 1\rangle_{\mathcal{C}',\mathcal{C}} = \lim_{j\to\infty}\langle w^{\psi_j}_{\varepsilon_j}, 1\rangle_{\mathcal{C}',\mathcal{C}} = \lim_{j\to\infty}\|\psi_j\|^2_{L^2(\mathbb{R}^d)}\,.$$

Hence, μ is bounded.

However, not all limit points are physically sensible. For example the bounded sequence $\psi_j(x) := \psi_0(x - j) \in L^2(\mathbb{R})$ has a limit point in $\mathcal{C}'$, some Banach-limit type functional, but the corresponding measure is zero. More generally, there are many continuous linear functionals on $\mathcal{C}$ which are zero on the (nondense) subset $C_0^\infty(\mathbb{R}^{2d})$.

Definition 1. *A sequence $\{\psi_j\}_{j\in\mathbb{N}}$ remains* localized in phase space *(with respect to $\{\varepsilon_j\}_{j\in\mathbb{N}}$), if it is compact at infinity, i.e.,*

$$\lim_{n\to\infty} \limsup_{j\to\infty} \int_{|x|\geq n} \mathrm{d}x \, |\psi_j(x)|^2 = 0\,,$$

and ε-oscillatory, i.e.,

$$\lim_{n\to\infty} \limsup_{j\to\infty} \frac{1}{\varepsilon_j^d} \int_{|p|\geq n} \mathrm{d}p \, |\widehat{\psi_j}(p/\varepsilon_j)|^2 = 0\,.$$

$\square$

Proposition 2. *Let $w_{\varepsilon_j}^{\psi_j} \overset{*}{\rightharpoonup} \mu$ in $\mathcal{C}'$ with $\{\psi_j\}_{j\in\mathbb{N}} \subset L^2(\mathbb{R}^d)$ bounded and localized in phase space, then μ has total mass*

$$\mu(\mathbb{R}^{2d}) = \lim_{j\to\infty} \|\psi_j\|_{L^2(\mathbb{R}^d)}^2\,, \tag{17.17}$$

and its marginals are given through the weak limits (in $\mathcal{M}_b^+$) of the quantum mechanical distributions, i.e., for all $a \in C_b^0(\mathbb{R}^d)$ one has

$$\int \mu(\mathrm{d}q, \mathrm{d}p) \, a(q) = \lim_{j\to\infty} \int \mathrm{d}q \, |\psi_j(q)|^2 \, a(q)\,, \tag{17.18}$$

$$\int \mu(\mathrm{d}q, \mathrm{d}p) \, a(p) = \lim_{j\to\infty} \varepsilon_j^{-d} \int \mathrm{d}p \, |\widehat{\psi_j}(p/\varepsilon_j)|^2 \, a(p)\,.$$

Proof. We start with the position marginal (17.18). Let $a \in C_b^\infty(\mathbb{R}^d)$ and let $\{a_n\}_{n\in\mathbb{N}} \subset C_0^\infty(\mathbb{R}^d)$ and $\{\chi_n\}_{n\in\mathbb{N}} \subset C_0^\infty(\mathbb{R}^d)$ satisfy $a_n(q) = a(q)$ and $\chi_n(p) = 1$ for $|q| \leq n$ resp. $|p| \leq n$. Then, by dominated convergence,

$$\int \mu(\mathrm{d}q, \mathrm{d}p) \, a(q) = \lim_{n\to\infty} \int \mu(\mathrm{d}q, \mathrm{d}p) \, a_n(q)\chi_n(p)$$

$$= \lim_{n\to\infty} \langle \mu, a_n\chi_n \rangle_{\mathcal{C}',\mathcal{C}} = \lim_{n\to\infty} \lim_{j\to\infty} \langle w_{\varepsilon_j}^{\psi_j}, a_n\chi_n \rangle_{\mathcal{C}',\mathcal{C}}$$

$$= \lim_{j\to\infty} \int \mathrm{d}q \, |\psi_j(q)|^2 \, a(q) + R\,,$$

where

$$\begin{aligned}
|R| &\leq \lim_{n\to\infty}\lim_{j\to\infty} |\langle w_{\varepsilon_j}^{\psi_j}, (a - a_n\chi_n)\rangle_{\mathcal{C}',\mathcal{C}}| \\
&\leq \lim_{n\to\infty}\lim_{j\to\infty} \left(|\langle \psi_j, (\widehat{a} - \widehat{a\chi_n})\,\psi_j\rangle| + |\langle \psi_j, (\widehat{a\chi_n} - \widehat{a_n\chi_n})\,\psi_j\rangle|\right) \\
&= \lim_{n\to\infty}\lim_{j\to\infty} \left(|\langle \psi_j, (\widehat{a} - \widehat{a\chi_n})\,\psi_j\rangle| + |\langle \psi_j, (\widehat{a\chi_n} - \widehat{a_n\chi_n})\,\psi_j\rangle|\right) \\
&\leq \lim_{n\to\infty}\lim_{j\to\infty} \left(\|\widehat{a}\psi_j\|\,\|(1 - \widehat{\chi}_n)\psi_j\| + \|(\widehat{a} - \widehat{a}_n)\psi_j\|\,\|\widehat{\chi}_n\,\psi_j\|\right) \\
&= 0\,.
\end{aligned}$$

For the last equality we used that $\{\psi_j\}$ is localized in phase space. In order to prove (17.18) also for $a \in C_{\mathrm{b}}^0$, note that we just proved that the right-hand side of (17.18) defines a measure. Hence, the result follows again by dominated convergence. The statements about the momentum marginal and the total mass follow analogously.

We now turn to the propagation of Wigner measures. As remarked in the introduction, a popular approach to the semiclassical limit of (17.4) is to determine the resulting transport equation for the Wigner measure associated with an ε-dependent initial condition

Corollary 2. *Let E_n be an isolated, nondegenerate Bloch band. Let μ_0 be the Wigner measure of a bounded sequence $\{\psi_{0,j}\}$ with $\psi_{0,j} \in \Pi_n^{\varepsilon_j} L^2(\mathbb{R}^d)$, i.e.,*
$$w_{\varepsilon_j}^{\psi_{0,j}} \overset{*}{\rightharpoonup} \mu_0 \in \mathcal{C}'.$$

Then the Wigner function $w_{\varepsilon_j}^{\psi_{t,j}}$ of the the time-evolved sequence

$$\psi_{t,j} := \mathrm{e}^{-\mathrm{i}H^{\varepsilon_j}t/\varepsilon_j}\psi_{0,j}$$

has the weak-$*$ limit $\mu_t \in \mathcal{C}'_{\mathrm{per}}$ given through

$$\mu_t = \mu_0 \circ \overline{\Phi}_0^{-t}\,. \tag{17.19}$$

In particular, μ_t is a positive bounded measure and solves the transport equation

$$\dot{\mu} + \nabla E_n(k - A(r)) \cdot \nabla_r\mu - \left(\nabla\phi(r) - \partial_l E_n(k - A(r))\nabla A_l(r)\right) \cdot \nabla_k\mu = 0$$

in the distributional sense.

Similar results were proved in [MMP, GMMP, BFPR] for the case of vanishing external potentials A and ϕ. For vanishing magnetic potential A but nonzero electric potential ϕ they follow from the results in [HST] or [BMP].

Proof (Proof of Corollary 2). According to Corollary 1 we have for $a \in \mathcal{C}_{\mathrm{per}}$ that

$$\left|\left\langle \left(w_{\varepsilon_j}^{\psi_{t,j}} - w_{\varepsilon_j}^{\psi_{0,j}} \circ \overline{\Phi}_0^{-t}\right), a\right\rangle_{\mathcal{C}',\mathcal{C}}\right| \leq \varepsilon_j\, C\, d_{\mathcal{C}}(a)\, \|\psi_{0,j}\|^2\,.$$

Taking the limit $j \to \infty$ on both sides yields the existence of the limit μ_t and at the same time (17.19). The transport equation for μ_t follows by taking a time-derivative in (17.19) and recalling that $\overline{\Phi}_0^t$ is the Hamiltonian flow of (17.2).

Acknowledgements. We are grateful to Caroline Lasser for helpful discussions on Wigner measures. This work was supported by the priority program "Analysis, Modeling and Simulation of Multiscale Problems" of the German Science Foundation (DFG).

References

[AsMe]   N.W. Ashcroft and N.D. Mermin, *Solid State Physics*, Saunders, New York, 1976.

[BFPR]   G. Bal, A. Fannjiang, G. Papanicolaou and L. Ryzhik, Radiative transport in a periodic structure, *J. Stat. Phys.* **95** (1999), 479–494.

[BMP]   P. Bechouche, N.J. Mauser and F. Poupaud, Semiclassical limit for the Schrödinger-Poisson equation in a crystal, *Comm. Pure Appl. Math.* **54** (2001), 851–890.

[Bu]   V. Buslaev, Semiclassical approximation for equations with periodic coefficients, *Russ. Math. Surveys* **42** (1987), 97–125.

[DGR]   M. Dimassi, J.-C. Guillot and J. Ralston, Semiclassical asymptotics in magnetic Bloch bands, *J. Phys. A* **35** (2002), 7597–7605.

[DiSj]   M. Dimassi and J. Sjöstrand, *Spectral Asymptotics in the Semi-Classical Limit*, London Mathematical Society Lecture Note Series 268, Cambridge University Press, 1999.

[GaAv]   O. Gat and J.E. Avron, Magnetic fingerprints of fractal spectra and the duality of Hofstadter models, *New J. Phys.* **5** (2003), 44.1–44.8.

[GMMP]   P. Gérard, P.A. Markowich, N.J. Mauser and F. Poupaud, Homogenization limits and Wigner transforms, *Commun. Pure Appl. Math.* **50** (1997), 323–380.

[GRT]   J.C. Guillot, J. Ralston and E. Trubowitz, Semi-classical asymptotics in solid state physics, *Commun. Math. Phys.* **116** (1988), 401–415.

[HST]   F.Hövermann, H. Spohn and S. Teufel. Semiclassical limit for the Schrödinger equation with a short scale periodic potential, *Commun. Math. Phys.* **215** (2001), 609–629.

[JNM]   T. Jungwirth, Q. Niu and A.H. MacDonald, Anomalous Hall effect in ferromagnetic semiconductors, *Phys. Rev. Lett.* **88** (2002), 207208.

[Ko]   W. Kohn, Theory of Bloch electrons in a magnetic field: The effective Hamiltonian, *Phys. Rev.* **115** (1959), 1460–1478.

[LiPa]   P.L. Lions and T. Paul, Sur les mesures de Wigner, *Revista Mathematica Iberoamericana* **9** (1993), 553–618.

[Lu]   J.M. Luttinger, The effect of a magnetic field on electrons in a periodic potential, *Phys. Rev.* **84** (1951), 814–817.

[MaNo]   A.Ya. Maltsev and S.P. Novikov, Topological phenomena in normal metals, *Physics - Uspekhi* **41** (1998), 231–239.

[MMP]   P.A. Markowich, N.J. Mauser and F. Poupaud, A Wigner-function theo-
           retic approach to (semi)-classical limits: electrons in a periodic potential,
           J. Math. Phys. **35** (1994), 1066–1094.

[MaRa]   J.E. Marsden and T.S. Ratiu. *Introduction to Mechanics and Symmetry*,
           Texts in Applied Mathematics Vol. 17, Springer Verlag, 1999.

[NeSo]   G. Nenciu and V. Sordoni, Semiclassical limit for multistate Klein-Gordon
           systems: Almost invariant subspaces and scattering theory, *Math. Phys.*
           Preprint, Archive mp_arc 01-36 (2001).

[PST$_1$] G. Panati, H. Spohn and S. Teufel, Space-adiabatic perturbation theory,
           Adv. Theor. Math. Phys. **7** (2003).

[PST$_2$] G. Panati, H. Spohn and S. Teufel, Space-adiabatic perturbation theory
           in quantum dynamics, *Phys. Rev. Lett.* **88** (2002), 250405.

[PST$_3$] G. Panati, H. Spohn and S. Teufel, Effective dynamics for Bloch electrons:
           Peierls substitution and beyond, *Commun. Math. Phys.* **242** (2003), 547–
           578.

[SuNi]   G. Sundaram and Q. Niu, Wave-packet dynamics in slowly perturbed
           crystals: Gradient corrections and Berry-phase effects, *Phys. Rev. B* **59**
           (1999), 14915–14925.

[Te]   S. Teufel, *Adiabatic Perturbation Theory in Auantum Dynamics*, Lecture
           Notes in Mathematics, Vol. 1821, Springer-Verlag, Berlin, Heidelberg,
           New York, 2003.

[TKNN]   D.J. Thouless, M. Kohomoto, M.P. Nightingale and M. den Nijs, Quan-
           tized Hall conductance in a two-dimensional periodic potential, *Phys.
           Rev. Lett.* **49** (1982), 405–408.

[Za]   J. Zak, Dynamics of electrons in solids in external fields, *Phys. Rev.* **168**
           (1968), 686–695.